U0155982

中华茶道图志

GRAPHICAL HISTORY OF CHINESE TEA CEREMONY

陕西历史博物馆

梁 子 。编著

陕西新华出版传媒集团

陕西人民出版社

图书在版编目（CIP）数据

中华茶道图志/梁子编著.—西安:陕西人民出
版社，2021.5
ISBN 978-7-224-14045-3

Ⅰ.①中… Ⅱ.①梁… Ⅲ.①茶道—中国—图集
Ⅳ.① TS971.21-64

中国版本图书馆 CIP 数据核字（2021）第 052204 号

责任编辑：韦禾毅
　　　　　王　辉
英文翻译：雷　钰
　　　　　黄晓红
　　　　　梁　敏

中华茶道图志

编 著 者　梁　子
出版发行　陕西新华出版传媒集团　陕西人民出版社
　　　　　（西安北大街 147 号　邮编：710003）
印　　刷　陕西龙山海天艺术印务有限公司
开　　本　898mm×1194mm　1/16
印　　张　35
字　　数　200 千字
版　　次　2021 年 5 月第 1 版
印　　次　2021 年 5 月第 1 次印刷
书　　号　ISBN 978-7-224-14045-3
定　　价　468.00 元

青山幽水流茶香　明月松风映真趣
——国画中的中华茶道

Water from Depth of Green Mountain Making Fragrant Tea Windy Pine under Bright Moon Reflecting Elegant Taste
—Chinese Tea Culture in Traditional Chinese Painting

　　清代江苏嘉定人陆廷灿（字幔亭）曾任崇安知县，负有进献武夷茶之责。为尽责，亦有情趣，其于采摘、蒸焙、汤候颇多观察研究，查阅有关茶书，分析见闻，遂成一代茶学重镇。其《续茶经》仿陆羽体例，对历代茶事均有论及。但其第九章《茶之图》在内容上与陆羽略有差异。从陆羽《茶经》看，所谓"十之图"是将《茶经》内容书写在素绢之上，张挂于饮茶之处，其实是茶馆设计的最原始思想。而陆廷灿则直接将有关茶道的绘画和茶器图案，记叙下来以为其主要内容，为人们

Lu Tingcan (namely Lu Manting), born in Jiading of Jiangsu Province of the Qing Dynasty, was the county magistrate of Chong'an, he was responsible for collecting Wuyi tea as royal court tribute as one of his daily official duties. He observed the procedures of tea-leaves picking, steaming and baking, tea making and made research on tea by consulting relevant books or written references due to not only his duty but also his own interest. Chong'an County turned into a tea science research center by his efforts. He wrote *Supplement of Tea Classic*, it is almost about all tea-related activities of the past dynasties by imitating Lu Yu's stylistic rules, but the chapter nine "Tea Catalog" is slightly different with *The Classic of Tea* by Lu Yu in content. The tenth chapter of *The Classic of Tea* corresponds to actually a wall map which can be copied onto a piece of silk and hung in the tea house to show the customers its content directly and instantly. This turns out to be one of the initial ideas on

认识茶道绘画打开更为真切明了的窗口。

陆廷灿《续茶经》之第九章《茶之图》所列历代图画名目有：

唐·张萱《烹茶仕女图》（见《宣和画谱》。）

宋·宣和御府藏 唐·周昉《烹茶图》一

宋·中兴馆阁藏 五代·陆滉《烹茶图》一

宋·周文矩《煎茶图》一、《火龙烹茶图》四

宋·李龙眠《虎阜采茶图》（见题跋）

范司理龙石藏 宋·刘松年绢画《卢全烹茶图》一卷，有元人跋十余家

王齐翰《陆羽煎茶图》（见王世懋《澹园画品》）

董逌《陆羽点茶图》，有跋。

元·钱舜举《陶学士雪夜煮茶图》（见詹景凤《东岗玄览》）

史石窗名文卿《煮茶图》，袁桷作《煮茶图诗序》

冯璧《东坡海南烹茶图并诗》

严氏《书画记》有杜柽居《茶经图》

汪珂玉《卢全烹茶图》（见《珊

tea house design. Lu Tingcan straightly put all tea-related paintings and pictures of tea ware together to be the major content of his book which of course provide a vivid collection for people who wish to know the tea-related paintings.

The names of tea-related paintings of the past dynasties listed in chapter nine "Tea Catalog" of Lu Tingcan's *Supplement of Tea Classic* are as follows:

"Tea Cooking Maiden" by Zhang Xuan of the Tang Dynasty from *Xuanhe Reign Edition Catalogue of Chinese Painting*

"Tea Cooking" by Zhou Fang of the Tang Dynasty preserved in royal palace of the Xuanhe Reign of the Northern Song Dynasty

"Tea Cooking" by Lu Huang of the Five Dynasties preserved in Zhongxing Imperial Library of the Song Dynasty

"Steaming Tea" and "Cooking Tea with Fiery Dragon" by Zhou Wenju of the Song Dynasty

See the preface and postscript of *Tea-leaves Picking* by Li Longmian of the Song Dynasty

Silk painting "Lu Tong Cooking Tea" by Liu Songnian of the Song Dynasty with postscripts of more than ten artists of the Yuan Dynasty

"Lu Yu Steaming Tea" by Wang Qihan from Wang Shimao's *Painting Appreciation*

"Lu Yu Brewing Tea" with postscript by Dong You

"Cooking Tea in Snowy Evening" by Qian Shunju of the Yuan Dynasty from *Seeing the Deep Reason* by Zhan Jingfeng

"Boiling Tea" by Shi Shichuang with poetic preface by Yuan Jue

"Su Shi Cooking Tea in Hainan with Poem" by Feng Bi

"The Classic of Tea" by Du Chengju from *Yan's Record*

瑚网》）

明·文徵明《烹茶图》

沈石田《醉茗图》，题云：
"酒边风月与谁同，阳羡春雷
醉耳聋。七碗便堪酬酩酊，任
渠高枕梦周公。"

沈石田《为吴匏庵写虎丘对
茶坐雨图》

陆包山《烹茶图》（见《渊
鉴斋书画谱》）

元·赵松雪《宫女啜茗图》（见
《渔洋诗话·刘孔和诗》）

上述茶道绘画，今能见者鲜。
除陆滉、董逌、冯璧、沈石田外，
其他画家作品有存。

今撰著《中华茶道图志》，
以与茶道相关紧密度论，分五个
层次。第一类，就是以品茗为主
体，不仅占画面主体位置（不一
定正中间），而且茶床、茶几、
茶器、茶具为主要道具，且以品
茗情思为画中主人的表情。比如
上述茶画外，有刘松年《磨茶图》，
郭纯《人物》、唐寅《事茗图》、
丁云鹏《煮茶图》、钱毂《竹亭
对棋图》、吕时敏《茗情琴意图》、
华喦《金屋春深图》、顾洛《小
倩小影图》等。第二类，不以品

of Calligraphy and Painting

"Lu Tong Cooking Tea" by Wang Keyu from *Rare Treasures*

"Cooking Tea" by Wen Zhengming of the Ming Dynasty

"Tasting Tea While Being Drunk" by Shen Shitian

"Drinking Tea at Huqiu When Raining in Name of Wu Pao'an" by Shen Shitian

"Tea Cooking" by Lu Baoshan from *Yuanjian Study Edition of Chart on Painting and Calligraphy*

"Court Maid Sipping Tea" by Zhao Songxue of the Yuan Dynasty from *Liu Konghe's Poem, Comment on Poetry*

It is rare for us to take a look at the above-mentioned tea-related paintings today, works of other artists have been preserved up to now except those by Lu Huang, Dong You, Feng Bi, and Shen Shitian.

The criterion based on the closeness of relation with tea ceremony for compiling this book *Graphical History of Chinese Tea Ceremony* consists of five classes. The first class includes paintings on tea wares, tea tasting, and emotional reaction in tea tasting such as the above-mentioned tea-related paintings and "Grinding Tea Leaves" by Liu Songnian, "Figures" by Guo Chun, "Self-entertainment by Drinking Tea" by Tang Yin, "Boiling Tea" by Ding Yunpeng, "Playing Chinese-chess in Pavilion among Bamboo" by Qian Gu, "Tea Tasting Accompanied by Music" by Lv Shimin, "Beauty's Spring Sorrow" by Hua Yan, "Portrait of Feng Xiaoqing" by Gu Luo, etc. The second class contains paintings on tea cooking as their partial content such as "Discussing Painting under Pine Tree by Creek" by Qiu Ying, "Talking about Old Days" by Tang Yin, "Zhong Kui and Beautiful Woman" by Ni Tian of the Qing Dynasty, "Virtuous Empresses and Imperial Concubines", and

茗为主题，但有明显的烹茶内容，其突出者有仇英《松溪论画图》、唐寅《西洲话旧》，清代倪田《钟馗仕女图》、冷枚《十宫词图》《邀月图》等。第三种情况属于雅集、豪族家居类的一些作品内容，如赵佶《文会图》、崔子忠《杏园宴集图》、杨晋《豪家佚乐图》、美国克利夫兰艺术馆藏叶芳林《九日行庵文宴图》。第四种情况是，候汤煮茗仅仅是角隅内容，例如西安考古所《柳岸货鱼图》、苏六朋《清平调图》及台湾故宫博物院《斟酒图》。第五种情况是取绘画适宜品茗的意境。以北京故宫《桃源仙境图》、西安博物院藏清人吴晨《高隐图》为典型，我们甚至看不到茶器，看不到饮茶的动作，但其意境却与品茗切题。

特别要指出的是，宋辽金元的出土壁画除耶律家族墓等有数幅大家公认的《备茶图》之外，大部分应该分别纳入上述第二、第三和第四种内容。还有，我们分别选取辽宁省博物馆和西安博物院的两幅仇英的《清明上河图》，也可以划入上述几个内容

"Drinking Invitation towards the Moon" by Leng Mei of the Qing Dynasty. The third class covers collection of works with refine taste and paintings on decorations and furniture of homes of the rich and powerful family such as "Literati Tea Party" by Zhao Ji, "Apricot Garden Tea Party" by Cui Zizhong, "Summer Enjoyment in Imperial Garden" by Yang Jin, and "Literati Tea Banquet" by Ye Fanglin preserved in the Cleveland Museum of Art in the US. The fourth class contains paintings with content of waiting for the boiling water or cooking tea at their corner parts such as "Fish Trading at Willow Bank" preserved in Xi'an Museum, "Li Bai Composing Yuefu Poem Summoned by Emperor Xuanzong of the Tang Dynasty" by Su Liupeng, and "Pouring Drink Picture" preserved in Taibei Palace Museum. The fifth class includes paintings with poetic environment suitable for tasting tea of which the representative works are "Wonderland in Peach Garden" preserved in Beijing Palace Museum and "Hermit" by Wu Chen of the Qing Dynasty preserved in Xi'an Museum. We even can't find any tea ware or tea tasting scene in these paintings, but its artistic conception comes to the point of tea drinking.

What need to be pointed out in particular is that among excavated tomb murals of the Liao, Song, Kin, and Yuan dynasties, most other murals of these dynasties should belong to the second, third, and fourth classes in addition to several murals excavated from Yelv and other family tombs having "Preparing Tea" pictures. We also selected two pictures with same name "Along the Bianhe River during the Qingming Festival" both by Qiu Ying but preserved respectively in Xi'an Museum showing us the early form of urban tea-house, and Liaoning Provincial Museum.

Here comes a problem. I tried hard however still could

之中。而且西安博物院的《清明上河图》为人们展现了都市茶馆的初期形态。

这就出现了一个问题，我无法囊括所有存世的以品茗论道为主体内容的茶道画，却将一部分山水画纳入书中，请同行们批评。过去我简单地认为，存世茶画就是我们以前曾经在图册、网络上见到过的茶道画。真正专心一意地整理时才发现，整理茶道画是一个复杂的工作。因为许多馆藏国画就从来没有面世过！我们经常见到的可能只是冰山一角！例如西安美术学院所藏国画品相很高，数量也不少。特别是吕焕成的《蕉阴品茶图》在茶道画中是不多见的。

虽然如此，我本人认为，中国茶道自唐宋以后，作为一种曾经十分高端的文化形态，渐趋衰微，甚至有人就说中国没有茶道。这可能有违历史实际。

"茶道"犹如"周礼"鲜见于馆阁文献，只好求之于文人画中。文人一直是中国茶道的中坚。饮茶作为一种生活习俗被视为开门七件事之最后一种，也就可有

not find and include all the preserved tea-culture paintings but put part of the landscape paintings with tea-culture content into this book wishing sincerely to know opinions from colleagues in this field. I simply thought in the past that the preserved tea-related paintings should be all of those we have seen in the atlas or on line. I found out that it was a complex work to clear up tea-culture-related paintings when I began to do it wholeheartedly because many of the traditional Chinese tea-culture-related paintings as part of museum collection have never been shown to the public, what can be seen in the museum exhibitions were only the tip of the iceberg. For instance, the traditional Chinese tea-culture-related paintings preserved in Xi'an Museum are of large quantity, high quality and good preserved condition especially "Tasting Tea Under the Shade of Banana Tree" by Lv Huancheng being a rare piece of work.

Even so, I think that Chinese tea ceremony which had been a very high-end cultural form since the Tang and Song dynasties have gradually declined, the statement of not having tea culture in China does not match with the historical fact.

The tea-culture description, just like the similar situation for the book *The Rites of Zhou*, is rarely seen in books and documents preserved in National Library, the only choice for me to find related references would be the literati paintings because scholars have been the backbone of the development of Chinese tea culture. Tea drinking had been put at the very end of the list on seven things of daily life which seems to be dispensable. What Emperor Gaozong (Hongli) of the Qing Dynasty said, "The state affairs can not be handled without an emperor, a gentleman can not live a single day without tea." It already had been lowered down to a shallow level of material life.

可无。至于弘历说"国不可一日无君，君不可一日无茶"，已经沦为浅层次的物质生活方面。

清朝直到民国，中国茶馆总要挂一个牌子，上书："莫谈国事"。这是将文人排除在高管队伍之外的必然结果。茶道，是一种道。道又是什么？道，就是一种人人得而效法的天下之公器。康熙乾隆、民国总统将治国权衡怀揣自己袖内，谁还敢学孔子，有老子在，谁敢说"修身齐家治国平天下"。

当郑燮的"难得糊涂"成为人们为人处世的座右铭时，追求"坚贞清白"最宜"精行俭德"的茶道何有立足之地！清代文化大家袁枚，对历史上的茶道清饮，应该有所体悟，他也只能说出的"味外味"这近乎佛教偈子的茶物语。

因此，从大的方面讲，茶道是一种文化形态，是对时代精神的折射。既然不能见诸文字，文人便诉诸丹青。在这些意境幽远、山青水静、茶香扑面而来的画图中，我们仍然能够体悟文人风骨和茶人情怀。茶事不仅是他们的日常生活，更是他们激扬文思的

Plates with written words "Don't Talk State Affairs!" could always be seen in the tea houses in China from the Qing Dynasty until the Republic of China. This was an inevitable result of excluding the literati from the senior executive group. The tea ceremony was a kind of doctrine. What was the doctrine then? It was a procedure accessible to everyone and easy to learn. Emperor Kangxi, Emperor Qianlong, and the presidents of the Republic of China exercised dictatorship involving governing the state with the theory of Taoism instead of that of Confucianism.

When Zheng Xie's view "Ignorance is a rare blessing" turned into people's life motto, there was no standing room for the core content of tea culture which was pursuing "faithful and innocence" and "refined behavior with virtue standard". Yuan Mei, a renowned scholar of the Qing Dynasty, had studied Chinese tea culture and had his own understanding of it, but what he expressed "subtext of tea tasting" was only a Buddhist-gatha-like conclusion on tea drinking.

Therefore, broadly speaking, the tea ceremony was a kind of cultural form and also a refraction of the spirit of the time. The literati presented their understanding and feeling towards tea drinking through paintings with drinking-tea scene in mountains surrounded by green trees and quiet brook. Drinking tea was not only part of their daily life but also a medium stimulated their inspiration and a Bodhimanda for their self morality cultivation.

Because of my limited capacity, I could not list all tea-culture related paintings from atlas and articles. What I could find out are as follows:

Tang and Song Dynasties (618-960)

"Xiao Yi Got 'Orchid Pavilion Preface' by Defrauding and

媒介、修身励志的道场。这些画都比较重视茶事活动的山水树木等环境的烘托与提炼。

本人不才，能力有限，可见图册及文章之有关茶画者不全，录述如下。

唐、五代（618—960）

阎立本（约601—673）《萧翼赚兰亭图》（辽宁省博物馆）

周昉（713—741）《调琴啜茗图》

孙位（约九世纪）《高逸图》（上海博物馆）

唐·无名氏《宫乐图》（台北故宫博物院）

五代·顾闳中（约910—980）《韩熙载夜宴图》

五代·周文矩《重屏会棋图》（北京故宫博物院）

宋、辽、夏、金、元（960—1368）

李公麟（1049—1106）《龙眠山庄图·延华洞》（台北故宫博物院）

无名氏《北齐校书图卷》

宋·王诜（1048—1104）《绣

Dedicated It to Emperor Taizong of Tang Dynasty" by Yan Liben (601−673) preserved in Liaoning Provincial Museum

"Fixing Gu Qin and Sipping Tea" by Zhou Fang (713−741)

"Seven Sages of the Bamboo Grove" by Sun Wei (About 9th Century) preserved in Shanghai Museum

"Royal Concubines' Banquet" by anonymous painter preserved in Taipei Palace Museum

"Evening Banquet Hosted by Han Xizai" by Gu Hongzhong (about 910−980) of the Five Dynasties

"Playing the Game of Go" by Zhou Wenju of the Five Dynasties preserved in Beijing Palace Museum

Song, Liao, Western Xia, Kin, and Yuan Dynasties (960–1368)

"Yanhua Cave, Sleeping−Dragon Mountain Villa" by Li Gonglin (1049−1106) preserved in Taibei Palace Museum

"Collating Five Classics and All kinds of Historical Records during Northern Qi Dynasty" by anonymous painter

"Meditation Before Mirror" by Wang Shen of the Song Dynasty

"Literati Tea Party" by Zhao Ji (1082−1135) namely emperor Huizong of the Southern Song Dynasty

"Calligraphy and Painting", *Comment on Poetry and Prose*

Murals on "Preparing Banquet, Music−accompanied Dance, Eating and Drinking at Banquet, Raising Children" of the Song Dynasty found at Heigou in Dengfeng, Henan Province

"Feast of Husband and Wife" found from tomb no.1 of the Song Dynasty at Baisha in Hebei Province

Brick−carving painting of the Song Dynasty from Guanyu's mausoleum in Luoyang City

Brick−carving painting "Waiting for Serving Tea" of the

栊晓镜图》，团扇

宋徽宗赵佶（1082—1135）《文
会图》

《鹤林玉露》——书画合璧

河南登封黑沟宋代壁画《备
宴图》《伎乐图》《宴饮图》《育
儿图》

河北白沙一号宋墓《开芳宴》

洛阳关林宋代砖雕画

山西长治市故漳村宋代砖雕
《奉茶恭候》

江参（南宋前期，师董源、
巨然）《千里江山图》（台北故
宫博物院）

南宋·苏汉臣（1094—1172）
《靓妆仕女图》

宋人物册页《斟酒图》（台北
故宫博物院）

宋·刘松年（1155—1218）《卢
仝邀茶图》

刘松年《缝纫·斟茶图》

刘松年《博古图》

刘松年《斗茶图》

刘松年《市井斗茶图》

刘松年《磨茶图》

刘松年《十八学士图》

刘松年《六人斗茶图》

李复题刘松年《卢仝烹茶图》

Song Dynasty found at Guzhang Village in Changzhi City of Shanxi Province

"Rivers and Mountains of Thousand Miles" by Jiang Can (lived in the early stage of the Southern Song Dynasty, learned painting technique from Dong Yuan and Juran) preserved in Taipei Palace Museum

"Ladies in Beautiful Make-up" by Su Hanchen (1094-1172) of the Southern Song Dynasty

"Pouring Liquor" from "Figure Picture Album" of the Song Dynasty preserved in Taipei Palace Museum

"Lu Tong Inviting Friend to Drink Tea" by Liu Songnian (1155-1218) of the Song Dynasty

"Lu Tong Cooking Tea" painted by Liu Songnian (1155-1218) and inscribed by Li Fu

"Sewing of the Pouring Tea" by Liu Songnian (1155-1218)

"Antiques" by Liu Songnian (1155-1218)

"Tea Game" by Liu Songnian (1155-1218)

"Tea Game at Marketplace" by Liu Songnian (1155-1218)

"Grinding Tea" by Liu Songnian (1155-1218)

"Eighteen Scholars" by Liu Songnian (1155-1218)

"Six-Man Tea Game" by Liu Songnian (1155-1218)

"Gathering of Celebrities at West Garden" by Ma Yuan

"Pouring Tea", "Antiques", "Tea Game", and "Grinding Tea" by Liu Songnian (1155-1218)

"Late Return from Spring Outing" by anonymous painter

"Preparing Tea of Five Hundred Lohans" by Yi Shao of the SouthPalace Museuemern Song Dynasty

"Cai Wenji's Captivity and Return from the Huns" by Chen Juzhong (served from 1201 to 1204 in Royal Academy during the Jiatai Reign of Emperor Ningzong) preserved in Taipei's

"Resigning Official Post and Returning Home" by He

马远《西园雅集图》

无名氏《春游晚归》

南宋·义绍《五百罗汉图·备茶》

陈居中（宁宗嘉泰年间为画院待诏）《文姬归汉图》（台北故宫博物院）

何澄（1223—？）《归庄图》（吉林省博物馆）

刘贯道（1258—1336）《消夏图》《梦蝶图》

钱选（1239—1289）《卢仝烹茶图》

赵孟頫（1254—1322）《斗茶图》

佚名《扁舟傲睨图轴》

朱德润（1294—1365）《秀野轩图卷》（北京故宫博物院）、《林下鸣琴图》（北京故宫博物院）

赵原《陆羽烹茶图》

佚名《楼阁图》（辽宁省博物馆）

佚名《竹林宴聚图轴》

佚名《舟中读书图》

盛懋（1340—1370）《清舸清啸图》（上海博物馆）

河北涿州元代壁画《备宴图》《云鹤线图》

陕西蒲城元代壁画《献酒图》

Cheng (1223-?) preserved in Jilin Provincial Museum

"Profound Scholar Enjoying Summer Coolness" and "Zhuangzhou Dreaming a Butterfly" by Liu Guandao (1258-1336)

"Lu Tong Cooking Tea" by Qian Xuan (1239-1289)

"Tea Game" by Zhao Mengfu (1254-1322)

"An Old Man on a Small Boat" by anonymous painter

"It is for Zhou Jing'an" and "Playing Gu Qin under Ancient Pine Trees" by Zhu Derun (1294-1365) preserved in Beijing Palace Museum

"Lu Yu Cooking Tea" by Zhao Yuan

"Pavilions" by anonymous painter preserved in Liaoning Provincial Museum

Hanging scroll "Gathering at Banquet in Bamboo Forest" by anonymous painter

"Reading in a Boat" by anonymous painter

"Preparing Banquet" and "Cloud and Crane" from Yuan Dynasty tomb mural at Zhuozhou in Hebei Province

"Offering Liquor" and "Singing and Dancing" from Yuan Dynasty tomb mural at Pucheng in Shaanxi Province

"Serving Tea" from Yuan Dynasty tomb mural at eastern suburbs of Xi'an City

"Husband and Wife Sitting Face-to-Face" found at Licheng District of Jinan City in Shandong Province

Murals on eastern, northern and southern walls of Liao-Dynasty tomb1 at Daxing in Beijing City

"Preparing Tea", "Drinking Liquor", and "Preparing Tea and Buddhist Sutras" from Liao-Dynasty tomb 1 at Xuanhua in Hebei Province, "Preparing Tea" on eastern wall of Liao-Dynasty tomb 7

"Preparing Tea" on eastern wall of front chamber of Liao-Dynasty tomb 7 at Xuanhua in Hebei Province,

《歌舞图》

西安东郊元代壁画《奉茶图》

山东济南历城区《夫妻对坐图》

北京大兴辽墓 M1 东壁、南壁壁画、东壁、北壁

河北省宣化辽墓 M1《备茶图》《喝酒图》《备茶、备经图》，M7 东壁《备茶图》

河北省宣化辽墓 M7 前室东壁《备茶图》，M7 后室东壁《持盏侍女图》，M10 前室东壁《备茶图》，M10《碾茶图》《煮水图》，M6 前室东壁《备茶图》，M5 后室东南壁《点茶图》，M5 后室西南壁《备茶图》，M2《奉茶进门图》，M2《备经图》《备茶图》，M4 后室东南壁《进茶图》，M4 后室西南壁《宴饮图》

辽宁法库县叶茂台辽肖义墓墓门旁西侧壁画《献食图》，东南壁《相迎图》

河北涿州辽代壁画墓东壁《宴乐图》

内蒙古滴水洞金墓壁画《点茶图》，金代（1115—1234）

山西大同市金代徐龟墓图

陕西甘泉金代壁画《夫妇宴

"Calyx-holding Maidservant" on eastern wall of rear chamber of Liao-Dynasty tomb 7; "Preparing Tea" on eastern wall of front chamber of Liao-Dynasty tomb 10, "Grinding Tea" and "Boiling Water" from Liao-Dynasty tomb 10; "Preparing Tea" on eastern wall of front chamber of Liao-Dynasty tomb 6; "Mixing Dust Tea and Boiling Water" on southeastern wall of rear chamber of tomb 5, "Preparing Tea" on southwestern wall of rear chamber of of tomb 5; "Serving Tea", "Preparing Buddhist Sutras" and "Preparing Tea" from tomb 2; "Offering Tea" on southwestern wall of rear chamber of of tomb 4, "Banquet" on southwestern wall of rear chamber of of tomb 4

"Offering Food" on western-side wall of Xiao Yi's tomb gate of the Liao Dynasty at Yemaotai in Faku County in Liaoning Province, "Greeting Guests" on southeastern wall

"Banquet and Music" on eastern wall of mural tomb of the Liao Dynasty at Zhuozhou in Hebei Province

"Mixing Dust Tea and Boiling Water" from Kin-Dynasty (1115-1234) mural tomb at Dishui Cave in Inner Mongolia

Mural from Xu Gui's tomb of the Kin Dynasty at Datong in Shanxi Province

"A Couple at the Feast" of Kin Dynasty mural at Ganquan in Shaanxi Province

"Husband and Wife Sitting Face-to-Face", "Figure of Ding Lan", "Figure of a Lady", and "Offering Tea and Cakes" at the left side of northern wall of Kin Dynasty mural tomb at Song Village in Tunliu of Shanxi Province

Ming Dynasty (1368-1644)

Hanging scroll "Woyun Thatched Cottage" by Lan Ying (1585-1664 of the Hongwu Reign) preserved in Beijing Palace Museum

饮图》

山西屯留宋村金代壁画墓北侧壁左侧《夫妻对坐图》《丁兰图》《夫人像》《奉茶点》

明代（1368—1644）

蓝瑛（1585—1664）《草堂卧云图卷》（北京故宫博物院）

谢环（生卒不详，洪武成名，永乐入宫，宣宗赐锦衣卫千户待遇）《杏园雅集图》（镇江博物馆）

杜琼（1396—1474）《南村别墅》十景图册之四阆杨楼（上海博物馆）、《友松图卷》

沈贞（1400—1482）《竹炉山房图》

吕文英（1421—1505）《货郎图·夏景》

沈周（1427—1509）《东庄图》（北京故宫博物院）、《虎丘送客图》（南京博物院）、《画山水》（台北故宫博物院）

郭纯《1370—1444》《人物》（台北故宫博物院）

文徵明（1470—1559）《真赏斋图》（上海博物馆）、《茶具十咏图轴》（北京故宫博物院）、《孝感图》（辽宁省博物馆）、《林

"Gathering in Apricot Garden" by Xie Huan (date of birth and death is unknown and famous in the Hongwu Reign, entered the royal palace in the Yongle Reign and was granted a treatment of Secret Police Organization in the Xuanzong Reign) preserved in Zhenjiang Museum

No. 4 of "Atlas of Ten Scenes of Nancun Villa", "Visiting Friend" by Du Qiong (1396–1474) preserved in Shanghai Museum

"Mountain House Surrounded by Bamboo Grove" by Shen Zhen (1400–1482)

"Street Vendor in Summer Day" by Lv Wenying (1421–1505)

"The Dongzhuang Villa" preserved in Beijing Palace Museum, "Seeing Off Guests by Playing Music" preserved in Nanjing Museum, and "Landscape" preserved in Taipei Palace Museum by Shen Chou (1427–1509)

"Figure" by Guo Chun (1370–1444) preserved in Taipei Palace Museum

"My Friend's Private House" preserved in Shanghai Museum, "Tea Tasting" preserved in Beijing Palace Museum, "Filial Piety" preserved in Liaoning Provincial Museum, "Entertaining Friend with Tea" preserved in Tianjin Art Museum, "Thatched Cottage" and "Tea Party at Foot of Mount Huishan" preserved in Beijing Palace Museum, "Judging Tea" and "Green Shadow Pavilion" by Wen Zhengming (1470–1559) preserved in Taipei Palace Museum

"Portrait of Lady by Imitating Tang-Dynasty Style", "Four Court Maids of Shu Palace", "Hermitage by Brook", "Talking About the Old Days", "Landscape", "Self-entertainment by Drinking Tea", and "Taogu Wrote a Poem for Qin Ruolan" by Tang Yin (1470–1524) preserved in Taipei Palace

榭煎茶图》（天津艺术博物馆）、《浒溪草堂图卷》、《惠山茶会图卷》（北京故宫博物院）、《品茶图轴》《影翠轩图轴》（台北故宫博物院）

唐寅（1470—1524）《仿唐人仕女图》《蜀宫宫妓图》《渔溪隐逸图》《西洲话旧》《画山水》《事茗图》《琴士图》《陶穀赠词图》（台北故宫博物院）

仇英（约1498—1552）《人物故事》、《写经换茶图》、《东林图》、《汉宫春晓》、《松亭试泉图轴》（台北故宫博物院）、《赤壁图》（辽宁省博物馆）、《清明上河图》（辽宁省博物馆）、《松溪论画图》（辽宁省博物馆）、《临溪水阁图页》（北京故宫博物馆）、《桃李园图》（日本知恩寺）、《桃源仙境图》（北京故宫博物馆）及《丽园庭深图》（西安美术学院）

谢时臣（1487—1567）《高人雅集图》

陆治（1496—1576）《竹泉试茗图轴》（台北故宫博物院）

王问（1497—1576）《煮茶图》（台北故宫博物院）

吴伟（1459—1508）《铁笛图

Museum

"The Character Story", "Zhao Mengfu Writing Buddhist Sutra for Tea Drink", "For Mr. Donglin", "Spring Morning in Palace", and "Pine Pavilion by Spring" by Qiu Ying (1498−1552) preserved in Taipei Palace Museum; "The Red Cliff", "Riverside Scene at Qingming Festival", and "Discussing Painting under Pine Tree by Creek" by Qiu Ying preserved in Liaoning Provincial Museum; "Pavilion Facing Stream" and "The Fairy Land of Peach Blossoms" by Qiu Ying preserved in Beijing Palace Museum; "Peaches and Plums Garden" preserved in Japanese Zhi−En Temple and "Deep West Garden" by Qiu Ying preserved in Xi'an Academy of Fine Arts

"Gathering of Profound Scholars" by Xie Shichen (1487−1567)

"Tasting Tea by Spring in Bamboo Grove" by Lu Zhi (1496−1576) preserved in Taipei Palace Museum

"Boiling Tea" by Wang Wen (1497−1576) preserved in Taipei Palace Museum

"Iron Flute" by Wu Wei (1459−1508) preserved in Shanghai Museum

"Playing Music under Plum Tree" by Du Jin (15th−16th Century)

"Seeing Guest Off at Wooden Door" by Zhou Chen (1487−1522) preserved in Nanjing Museum

"Boiling Spring Water" and "Playing the Game of Go in Bamboo Pavilion" by Qian Gu (1508−1572) preserved in Liaoning Provincial Museum, "Profound Scholar in Bamboo Grove by Stream" preserved in Xi'an Museum and "Enjoying Coolness in Summer" by Qian Gu preserved in Xi'an Academy of Fine Arts

"Smaller Version of Painting and Postscript by Imitating

卷》（上海博物馆）

杜堇（15—16世纪初）《梅下横琴图》

周臣（1487—1522）《柴门送客图》（南京博物院）

钱毂（1508—1572）《惠山煮泉图》、《竹亭对棋图》（辽宁省博物馆）、《竹溪高士图》（西安博物院）及《桐溪消夏图》（西安美术学院）

董其昌（1555—1636）《仿宋元人缩本画及跋》册页

明·戴进（1388—1462）《春酣图》（台北故宫博物院藏）

尤求《红拂图》（北京故宫博物院）

佚名《千秋绝艳图》

佚名《入跸图》

刘俊《雪夜访普图》

钱贡（16—17世纪初）万历十四年（1586）《山水间》、万历四十年（1612）《鱼乐图》《岁寒图》《太平春色》

项元汴（1525—1590）《梵林图卷》（南京博物院）

李士达［16—17世纪初，万历二年（1574）进士］《坐听松风图》（台北故宫博物院）

Song-Yuan-dynasties Style" by Dong Qichang (1555-1636)

"Villagers' Spring Sacrificial Feast in the Suburb" by Dai jin (1388-1462) of the Ming Dynasty preserved in Taipei Palace Museum

"Singing Girl" by You Qiu (who lived between the Jiajing Reign and the Wanli Reign of the Ming Dynasty) preserved in Beijing Palace Museum

"Noble Women in Past Dynasties" by anonymous painter

"Emperor Shenzong Coming Back to the Palace by Boat after Visit to Ancestral Tomb" by anonymous painter

"Zhao Kuangyin-Taizu Emperor of the Song Dynasty Paying a Visit to Prime Minister Zhao Pu in Snowy Evening" by Liu Jun

"Landscape", "Swimming Fish", "Cold Season", and "The Spring Scenery in Peaceful Times" by Qian Gong (16th-17th Century)

"Buddhist Temple" by Xiang Yuanbian (1525-1590) preserved in Nanjing Museum

"Sitting and Listening to In-the-Pines Whispering Wind" by Li Shida (16th- early 17th Century, selected as a successful candidate in the highest imperial examinations in 1574, 2nd year of the Wanli Reign) preserved in Taipei Palace Museum

"Enjoying the View of Snow of Activities While Staying at Home Idle" and "Boiling Tea of Activities While Staying at Home Idle" by Sun Kehong (1532-1611) preserved in Taipei Palace Museum

"Emperor Xuanyuan Asking the Way of Nature" by Shi Rui

"Boiling Tea" and hanging scroll "Filtering Liquor" by Ding Yunpeng (1547-1618) preserved in Shanghai Museum

孙克弘（1532—1611）《消闲清课图卷·赏雪》《消闲清课图卷·煮茗》（台北故宫博物院）

石芮《轩辕问道图》

丁云鹏（1547—1618）《煮茶图》《滤酒图轴》（上海博物馆）

袁尚统（1570—1661）《岁朝图》（苏州博物馆《明清书画图册》）

项圣谟（1597—1658）《谢彬松涛散仙图轴》

吴彬（1573—1643）《文士雅集图》（西安美术学院）

崔子忠（1574—1644）《杏园宴集图》

陈洪绶（1599—1652）《品茶图》《授徒图》（绢本设色，美国加州大学美术馆）

余令《煮茶闻道图》（西安美术学院）

佚名《云山深处听琴图》（西安美术学院）

明·佚名《南都繁会图卷》

佚名《西园雅集图》（西安美术学院）

清

髡残（1612—1692）《幽栖图》

周洽（1625—1700）《竹溪春

"The Beginning of a Year" by Yuan Shangtong (1570–1661) from "Atlas of Ming–and–Qing Dynasties Calligraphy and Paintings" preserved in Suzhou Museum

Hanging scroll "The Old Man Walking between Pine Trees" by Xiang Shengmo (1597–1658) and Xie Bin

"Gathering of Literati" by Wu Bin (1573–1643) preserved in Xi'an Academy of Fine Arts

"Gathering at Banquet in Apricot Garden" by Cui Zizhong (1574–1644)

"Tasting Tea" and "Teaching Skill to Female Apprentices" (silk scroll with ink and color) by Chen Hongshou (1599–1652) preserved in Art Gallery of the University of California

"Boiling Tea and Discussing the Way of Nature" by anonymous painter (Lang Yuling from Dingzhou) preserved in Archaeological Institute of Xi'an Museum

"Listening to Music in Cloudy Mountain" by anonymous painter preserved in Xi'an Academy of Fine Arts

"Nanjing City in the Ming Dynasty" by Ming–Dynasty People

"Gathering of Literati in West Garden" by anonymous painter preserved in Xi'an Academy of Fine Arts

Qing Dynasty

"Dwelling in Mountain Cottage" by Kun Can (1612–1692)

"Spring Scenery of Bamboo Creek" by Zhou Qia (1625–1700) preserved in Xi'an Academy of Fine Arts

"Tasting Tea in the Shadow of Banana Tree" by Lv Huancheng (1630–1705) preserved in Xi'an Academy of Fine Arts

"Cooking Tea with Spring Water", "Mountain Cottage", and "Landscape by Imitating Xu Daoning's Style" by Wang

昼图》（西安美术学院）

吕焕成（1630—1705）《蕉阴品茶图》（西安美术学院）

王翚（1632—1717）《石泉试茗轴》（台北故宫博物院）、《一梧轩图轴》（台北故宫博物院）、《临许道宁山水轴》（台北故宫博物院）

吴历（1632—？）《云白山青图》

吴宏（生卒年不详）《柘溪草堂图轴》（南京博物院）

杨晋（1644—1728）《豪家佚乐图》（南京博物院）

禹之鼎（1647—1716）《幽篁坐啸图》

王树毂（1649—？）《四友图轴》（北京故宫博物院）

郎廷极（1663—1715）《郎廷极行乐图》（自题名《茗情琴意图》）（青岛市博物馆）

颜峄（1666—？）《江楼对弈图轴》（广州美术馆）

陈枚（？—1864），上海人，所绘《月曼清游图》之《秋桐聚戏图》（北京故宫博物院）

佚名　《刻丝夜宴桃李园》（原为宫廷制造，辽宁博物馆）

王概（生卒不详，康熙时作品较多）《秋帆旷揽图》（上海

Hui (1632-1717) preserved in Taipei Palace Museum

"Landscape" by Wu Li (1632-?)

"Cottage at Zhexi" by Wu Hong (whose date of birth and death is unknown）preserved in Nanjing Museum

"Noblewomen Amusing Themselves" by Yang Jin (1644-1728) preserved in Nanjing Museum

"Wang Shizhen Touching Chords of Guqin" by Yu Zhiding (1647-1716)

"Plum Blossoms, Orchid, Bamboo, and Chrysanthemum" by Wang Shugu (1649-?) preserved in Beijing Palace Museum

"Listening to Music While Tasting Tea" by Lang Tingji (1663-1715) preserved in Qingdao Municipal Museum

"Playing the Game of Go" by Yan Yi preserved in Guangzhou Municipal Art Gallery

"Palace Ladies Enjoying Life" by Chen Mei (?-1864) from Shanghai preserved in Beijing Palace Museum

"Evening Banquet in Peaches and Plums Garden" by anonymous painter preserved in Liaoning Provincial Museum

"Autumn Scene of the Countryside around Nanjing" preserved in Shanghai Museum, "Landscape in Autumn" preserved in Tianjin Art Museum and "Autumn Landscape" preserved in Zhejiang Provincial Museum by Wang Gai (whose date of birth and death is unknown, most of his works were created during the Kangxi Reign)

"Drinking Invitation towards the Moon" created in 1739 by Leng Mei (about1669-1742) preserved in Xi'an Academy of Fine Arts

"Spring Dinner in Peaches and Plums Garden" by Huang Shen (1687-1768)

"Emperor Qianlong Appreciating Painting" by Lang

博物馆）、《秋景山水图轴》（天津艺术博物馆）、《秋山喜客图卷》（浙江省博物馆）

冷枚（约 1669—1742）1739 年作《邀月图》（西安美术学院）

黄慎（1687—1768）《春夜宴桃李园图》

郎世宁（1688—1766）《弘历观画图像》（北京故宫博物院）

姚文翰（18 世纪）《四序图》

华嵒（1682—1756）《金屋春深图》（广东省博物馆）、《金谷园图轴》（上海博物馆）、《竹溪六逸图》（水墨设色，日本中西文之藏）

顾洛（1763—约 1837）《小青小影图》

金廷标（？—1767，1760 年入宫供奉）《仙舟笛韵》《品泉图》

潘承桂（生卒年不详）《芦湖赏月图》《琴瑟和鸣图》

苏六朋（1791—1862）《清平调图》《游僧晚归图》《秋赏图》

王素（1794—1877）《高士论诗图》（西安博物院）

费丹旭（1802—1856）《倚栏图卷》《姚燮纤绮图》《听秋啜茗图》（浙江省博物馆）

真然（1816—1884）《淮阴垂

Shining (1688－1766) preserved in Beijing Palace Museum

"Palace Ladies' Life of Leisure in Four Seasons" by Yao Wenhan (the 18th Century)

"Palace Lady" preserved in Guangdong Provincial Museum, "Listening to Music in Jingu Garden" preserved in Shanghai Museum, and "Six Profound Scholars Tasting Tea in Bamboo Grove" (ink and color）preserved in Japan by Hua Yan (1682－1756)

"Portrait of Xiaoqing" by Gu Luo (1763－1837)

"The Charm of Flute Music" and "Tasting Spring Water" by Jin Tingbiao (who lived between the Qianlong Reign and the Yongzheng Reign, entered into the royal palace in 1760)

"Appreciating the Moon on Luhu Lake" and "The Lute and Psaltery Are in Harmony" by Pan Chenggui (whose date of birth and death is unknown) preserved in Xi'an Academy of Fine Arts

"Li Bai Composing Poem for Emperor Xuanzong", "Traveling Monk Coming Back Late" and "Appreciating Autumn Scenery" created in 1845 by Su Liupeng (1791－1862) preserved in Xi'an Academy of Fine Arts

"Profound Scholar Discussing Poem Composition" by Wang Su (1794－1877) preserved in Archaeological Institute of Xi'an Museum

"Leaning against Pavilion Railing", "Yao Xie's Daily Life", and "Listening to the Sound of Autumn while Sipping Tea" by Fei Danxu (1802－1856) preserved in Zhejiang Provincial Museum

"Fishing at Huaiyin" by Zhen Ran (1816－1884) preserved in Xi'an Academy of Fine Arts

"Happy Drinking at Feast" and "Carrying Guqin to Visit Friend" by Sha Fu (1831－1906) preserved in Xi'an Academy

钓图》（西安美术学院）

沙馥（1831—1906）《琼筵飞觞图》《携琴访友图》（均藏于西安美术学院）

张筠（生卒年不详），近代画家，浙江秀水（今嘉兴）人，光绪九年（1883）翰林，后任四川主考官。精诗文工书画，1861年作《说书纳凉图》（西安美术学院）

钱慧安（1833—1911）《烹茶洗砚图》（上海博物馆）、《簪花图》、《人物花卉四条屏·柳下背书图》（西安考古所）

任薰（1835—1893）《人物》（台北故宫博物院）、《临泉调琴图》（西安美术学院）

吉朝（生卒年不详），1899年作《灵山积玉图》（西安美术学院）

吴昌硕（1844—1927）《品茗图》（上海朵云轩）

倪田（1855—1919）《钟馗仕女图》（徐悲鸿纪念馆）

黄钺徵（生卒年不详）《黄钺徵符康阜册·晓雾烹茶图》（台北故宫博物院）

陈字（1634—？）《青山红树图轴》（北京故宫博物院）

王树毂（1649—？）《四友图

of Fine Arts

"Telling Story While Enjoying Coolness" created in 1861 by Zhang Yun (whose date of birth and death is unknown, a painter of modern times, born in Xiushui, today's Jiaxing of Zhejiang Province, appointed as an academician in 1883, 9th year of Guangxu Reignand later chief examiner in Sichuan Province, good at poetry, calligraphy and painting) preserved in Xi'an Academy of Fine Arts

"Cooking Tea and Washing Ink-stone", "Wearing Flower as Hairpin", and "Reciting A Lesson under the Willow Tree" by Qian Hui'an (1833—1911) preserved in Archaeological Institute of Xi'an Museum

"Figure" preserved in Taipei Palace Museum and "Fixing Guqin by the Creek" preserved in Xi'an Academy of Fine Arts by Ren Xun (1835—1893)

"Landscape" created in 1899 by Ji Chao (whose date of birth and death is unknown) preserved in Xi'an Academy of Fine Arts

"Tasting Tea" by Wu Changshuo (1844—1927) preserved in Duoyunxuan Publishing Company in Shanghai

"Zhongkui and Noble Lady" by Ni Tian (1855—1919) preserved in Xu Beihong Memorial

"Cooking Tea in the Morning Fog" by Huang Yuezheng (whose date of birth and death is unknown) preserved in Taipei Palace Museum

"Green Hills and Mangrove" by Chen Zi (1634—?) preserved in Beijing Palace Museum

"Plum Blossoms, Orchid, Bamboo, and Chrysanthemum" by Wang Shugu(1649—?) preserved in Beijing Palace Museum

"Servants at Tea Competition" by Wang Chengpei (?—1805)

轴》（北京故宫博物院）

汪承霈（？—1805）《群仙集祝图》

袁耀（？—1788）《潇湘烟雨图轴》（安徽博物院）

吴晨（生卒年不详）《山水册》（西安考古所）

潘振镛（1852—1921）《天街夜色》（西安美术学院）

"Landscape of Hunan" by Yuan Yao (?−1788, lived in the late Kangxi Reign) preserved in Anhui Provincial Museum

"Appreciating the Moon on Luhu Lake" created in 1896 and "The Lute and Psaltery Are in Harmony" in 1899 by Pan Chenggui (whose date of birth and death is unknown)

"Landscape" by Wu Chen (whose date of birth and death is unknown) preserved in Archaeological Institute of Xi'an Museum

"Night Sky" by Pan Zhenyong (1852−1921) preserved in Xi'an Academy of Fine Arts

目录
Table of Contents

引言
Introduction

第一章 陆羽湖州茶文化圈与茶道文化的诞生
Chapter One Tea Culture Circle in Huzhou and Birth of Tea Ceremony Culture

第二章 宋、辽、夏、金、元时代茶文化及茶画制作
Chapter Two Tea Culture and Tea-Related Painting Creation of the Song, Liao, Xia, Kin, and Yuan Dynasties

第三章　明代茶道文化整体评价与茶画创作

Chapter Three　Overall Evaluation of Tea Culture and Tea-Related Painting Creation of the Ming Dynasty

第四章　清代茶文化的整体风貌

Chapter Four General Feature of Tea Culture in the Qing Dynasty

跋
Postscript

中国茶道文化概要和茶道模式的历史演变
Summary of Chinese Tea Culture and Historical Evolution of Tea Ceremony

陈文华《茶文化学》将茶文化学分为物质、精神、制度和心态四个层次。而他的论述以狭义茶文化，即茶的精神文化为主。我们以为应该统称为中国茶道文化，包括这四个层次的科技知识的积累和精神文化的积淀。茶树的栽培改良和茶叶的加工制作，不同于花卉蔬菜，也不同于地质矿产和物理等自然学科。不论茶树栽培者、茶叶加工者运用多么高端的现代理论和设备进行种茶和加工，其最终目的是为人类提供一杯好茶。其所有工作是以此目标为中心。茶道是什么?就是用心烹出一杯好茶。一杯好茶既是结果也是目的。一杯好茶需要一壶好水、一手精湛的烹茶技艺、一套好的茶具、一袋好的茶叶……一袋

Chen Wenhua divided Chinese tea culture in his book *Tea Culturology of China* into four levels: material, spiritual, institutional, and mental ones. And that he discussed Chinese tea culturology in the narrow sense. We think it should be called by a joint name "Chinese tea culture" containing the accumulation of scientific and technological knowledge and the sedimentary deposits of cultural heritage on four levels. Tea trees plantation and improvement and tea manufacture are different from that of flowers and vegetables, also different from that of natural sciences such as geology, mineral resources, and physics. The ultimate aim of those tea tree planters and tea producers is to provide high-quality tea for all of us no matter what modern equipment have been used. What is tea culture then? The core content of tea culture is to cook a nice cup of tea wholeheartedly which is not only the result but also the purpose of it. A pot of clear water, skillful technique, a good tea set, and a bag of high-quality tea leaves make a cup of good tea. A bag of high-quality tea leaves requires good soil, suitable fertilizer and

好茶叶需要好土壤好肥料好的加工设备……说到底，要有一颗好心。这就是茶之心，是进行过修炼而成的茶人之心。这颗好心是用中国茶道理念武装起来的善良心、博爱心、智慧心、精进心。这颗心应追求清雅，向往和谐，崇尚德俭，健行精进。

在长达三千多年以上的茶事历史中，国人经过了采集时代的食用、文明时代的药用和封建时代的饮用。就食用而言，可以生嚼，也可以炒熟或煮熟吃。就饮用而言，又经过了汉代至唐代中期的晒青、蒸青，后釜中煎煮。煮之前先要将团茶或饼茶敲碎碾细。煮茶时，还要加入胡椒、生姜和盐。陆羽《茶经》推崇制作蒸青团茶，反对煮茶时加上其他佐料，但保留了加盐。对鉴水、鉴茶、注水、茶器制作、茶叶保养和茶室布置都提出了具体要求，并指出："茶宜精行俭德之人。"这是有关中国茶道精神的最早表示。有人依据刘禹锡《西山兰若茶歌》判断，中晚唐以后文人士大夫和高僧名道开始采用炒青法——将绿茶现摘现炒现煮，这里就是散茶现煮。但也有人认为诗中的旋采旋炒，指的是煮饮新鲜茶叶而不经过晒、晾、蒸等工序。从《旧唐书·地理志》记载土贡中，金州（今陕西安康

processing equipment, more importantly a loving-tea heart namely a tea-man heart full of kindness, philanthropism, wisdom, and turning towards good derived from long-term cultivation. A tea-man heart longs for elegance, harmony, morality, living-with-less and turning towards good.

In more than three thousand years of tea-related history, Chinese people had eaten tea in the food-collecting era, taken tea as medicine in the civilized era and as a kind of drink in the feudal era. Tea leaves can be chewed directly, eaten after being cooked or boiled in terms of food; had been dried in the sun or steamed then boiled in the cauldron from the Han Dynasty to middle Tang Dynasty in terms of a drink.In fact, it had been grinded into fine tea dust before cooked with pepper, ginger and salt. Lu Yu praised highly in *The Classic of Tea* making steamed green cake-shaped tea adding only salt, opposed against cooking tea with other seasonings. Lu Yu put forward the detailed requirements for selecting water, distinguishing tea leaves, pouring water, making tea set, tea leaves maintenance, and tea-house arrangement, then pointed out that, "Tea man should always keep traditional Chinese moral spirit." This is the earliest expression on connotation of Chinese tea culture. Scholars, bureaucrats, eminent monks, and famed taoists were believed to cook fresh tea leaves instantly and drink it after the middle and late Tang Dynasty according to the poem "Cooking Fresh Tea Leaves for Entertaining Friend" by Liu Yuxi. But some people thought that related lines in this poem referred to boil fresh tea leaves without being dried in the sun, cooled in the air and steamed. Concerned records in the "Geography" of *Old Tang Book* about the royal contributes of fresh tea buds from Ankang City of Shaanxi Province and *The Record of a Pilgrimage to China in Search of the Law* about the loose tea given to Yuanren and his fellow men as a

市）贡"茶芽"；《入唐求法巡礼行记》又记载圆仁一行得到大唐僧俗友人赠送散茶，这样就表明：至迟在九世纪前期，大唐已经饮用散茶。

晚唐苏廙《十六汤品》则告诉人们：晚唐已经采用"淹茶法"，以汤"沃"焉，实际就是点茶。有学者认为法门寺出土的琉璃茶托茶盏也佐证了点茶法的存在。

沃茶，也叫淹茶法。王潮生先生认为"淹"应该读yǎn，其原意是一种半卧半起的病。因为陆羽认为这种茶是一种半生茶，因而定名为淹茶，是一种病态的茶法。还有一种解说为草庵茶，言其自然简易，这更是一种望文生义。我们认淹茶之淹，本意更接近腌，即先民把菜放在坛子罐子或瓮中加上盐浸泡，类似泡菜法。《茶经·六之饮》云：饮有粗茶、散茶、末茶和饼茶，"乃斫、乃熬、乃炀、乃舂，贮于瓶缶之中，以汤沃焉，谓之淹茶。或用葱、姜、枣、橘皮、茱萸、薄荷等，煮之百沸，或扬令滑（清），或煮去沫，斯沟渠间弃水耳，而习俗不已"。

实际上陆羽已经向我们介绍了三种吃茶法：（1）将四种茶砍斫、炒干、研磨后用瓶缶贮茶，用开水冲泡；（2）用瓶缶熬煮；（3）他提倡的用缶烹茶。除了少量盐之外，不

gift from Tang-Dynasty royal court show that loose tea had been used within the Tang Empire no later than early 9th century.

Su Yi of the late Tang Dynasty wrote in *Sixteen Tea Liquors* that the method of making tea by pouring boiling water into cup with grinded tea dust was adopted in the late Tang Dynasty. Some scholars agreed to this view after seeing colored-glaze tea calyx and its saucer unearthed at Famen Temple.

The making-tea method is called "Wo" also "Yan" in Chinese. Mr. Wang Chaosheng thinks that the word "Yan" means a pathological state with semi-reclining life style, while Lu Yu called it "Yan" meaning tea made by means of a pathological method because he thought this kind of tea was only a half-cooked one, another name for such tea "Cao'an" referring to "thatched hut" meant natural and simple tea. We think that the key word of the name of this sort of tea "Yan" means almost the same as the way of dipping fresh vegetable in salty water in a glazed pottery jar or urn like kimchi prepared every year by Koreans. The "Sixth Section Tea Drinking" of *The Classic of Tea* recorded that there were coarse tea, loose tea, dust tea, cake-shaped tea according to the shape of tea, and tea was chopped, boiled, baked, pounded in the early stage of tea making, then was put into pot or cup, dipped into boiling water, there got the "Yan" tea. Some people put even onion, ginger, jujube, orange peel, dogwood, mint, etc. in the tea which could not be regarded as a real tea but obsoleting water which need to be poured into the ditch, still this was a custom in some places.

Lu Yu actually introduced to us three kinds of tea-making method in *The Classic of Tea*. 1. Chop, stir-fry, grind coarse tea, loose tea, dust tea, and cake-shaped tea, put it into pottery vase or jar for storage, make tea with boiling water. 2. Boil tea with pottery vase or jar. 3. Cook tea

加其他调料。我们可以这样想象：砍斫带着茶树枝条细木的晒青粗放的原叶茶，砍碎切小，这种茶必须熬煮，因为带着枝杈。散茶——没有树枝的茶叶，或晒青或蒸青，这种茶用来熬煮。末茶，也就是前两种剩余的茶叶粉末，或故意揉碎的末茶，事先炒一下。炀的本意就是熔化，也就是烘焙，再直接用开水冲泡；饼茶必须砸碎，加以研磨，就是舂。放在木砧板上敲碎，或在石臼中舂碎。可以冲泡——沃。沃，就是用水浇覆。陆羽提倡缶中烹茶，反对以上提到的其他两种吃茶法，但是这些方法已经延续了400多年，他没办法，只能说"习俗而已"。

中晚唐时期的点茶道日益普遍化。福建茶人每年在采茶季节都要进行点茶比赛，当地人称之为"茗战"。可能在晚唐以后，中国的气候发生变化，气温明显下降，于是中国主要贡茶区从苏浙向福建南移。建州茶成为北宋政府的贡茶基地。两宋以点茶为主要茶道形式。宋徽宗的《大观茶论》和蔡襄《茶录》成为宋代茶道的经典型文献。"祛襟涤滞，致清导和；冲淡简洁，韵高致静"是他们对中国茶道精神的阐述。纯香料和花香入茶成为饮茶习俗的新发展。

with pottery vessels adding a small amount of salt, no other seasonings.Let us imagine this： chop dried tea leaves even having tea tree branches with it, then boil it because of its fine tea tree branches; stew loose tea leaves without branch having been dried in the sun or steamed; would better bake dust tea in advance, then make with boiling water directly; cake-shaped tea must to be smashed first, pounded on wooden cutting board or in stone mortar, finally made with boiling water. Lu Yu advocates to cook tea in pottery vessel rather than other methods, but these methods have lasted for over 400 years. Lu Yu could do nothing about it but saying "custom".

Making tea with boiling water became more and more popular in the middle and late Tang Dynasty. Tea man in Fujian province organized annual tea-making competition called by local people "Tea Battle" in the tea-plucking season. China's major tea producing area moved southward from Jiangsu Province and Zhejiang Province to Fujian Province after the late Tang Dynasty due to the climate changing which led to the drop of the temperature drastically. Jianzhou tea thus became the royal-tribute tea in the Northern Song Dynasty. Making tea with boiling water had been the main method during the Northern Song and Southern Song Dynasties. *Discussion on Tea* by Emperor Huizong of the Southern Song Dynasty and *Tea Records* by Cai Xiang of the Northern Song Dynasty turned out to be the classical literature of the tea culture of the Song Dynasty. Their understanding and illustration on core connotation of Chinese tea culture is "Tea helps people to get clear and pure mind, harmonious relation with nature, leisurely and comfortable state, and rhythmic tranquility." Adding pure spice and flower petal into tea as its ingredients became another tea-related custom.

明代早期延续了宋代主流茶道模式点茶法。本着勤俭简朴的治国原则，洪武皇帝废除凤团龙饼，提倡散茶冲饮。其第十七子朱权（1378—1448）的《茶谱》是明朝茶道文化代表。"助诗兴、伏睡魔、倍清谈"是他对茶道文化的理解。他对茶炉、茶灶、茶磨、茶罗等茶器尺寸提出具体的设计要求。明代中期散茶煎煮逐渐流行，专门用以冲泡散茶的紫砂壶开始登上中国茶文化的历史舞台。周高起（？—1645）的《阳羡茗壶系》是第一部谈论宜兴紫砂陶的著作。

成书于1735年（雍正十三年）前的陆廷灿的《续茶经》，依照陆羽《茶经》结构和体例对中国茶文化进行了梳理。其中的《茶之图》将历代烹茶图悉数记录，也对十二茶具图悉数罗列。附录部分还历数唐德宗以来的有关茶叶的法规制度。在散茶冲泡中，因偶然因素人们发现了半发酵的大红袍（乌龙茶）味道别有韵味而采用发酵法制茶，浙江龙井茶因乾隆皇帝南巡时啜饮而名满天下。而乌龙茶的出现，距离全发酵的红茶仅仅一步之遥。17世纪福建崇安小种红茶出现，随后发明功夫红茶，进而推向全国。

明清以来，为适应茶叶外销，出现紧压茶，因渥堆发酵，颜色变为黑

Making tea with boiling water had been the mainstream tea-making mode in the early Ming Dynasty. "Coiled-Dragon-and-Phoenix Pie" tea was abolished by the Taizu Emperor of the Ming Dynasty based on state-governing principle of simple, hardworking and thrifty, loose tea was advocated to be used. Zhu Quan (1378-1448), the 17th son of the Taizu Emperor wrote a representative book of Ming-Dynasty tea culture, *Tea Manual*, illustrated his own understanding on tea culture: inspire poetic creation, prostrate desire to sleep and create joyful atmosphere for friends' talking. He proposed the specific size for water boilor, tea stove, tea millstone and tea sifter. Loose tea cooking led the fashion in the Mid-Ming Dynasty, purple clay tea pot specially for making loose tea started to board the stage of the history of Chinese tea culture. *Purple Clay Teapot* by Zhou Gaoqi (?-1645) is the first book talking about the Yixing purple clay ware.

Sequel to the Classics of Tea on Chinese tea culture by Lu Tingcan finished in 1735 (the year of Yimao of Yongzheng Reign of the Qing Dynasty) was written according to the structure and style of Lu Yu's *The Classic of Tea*. "Chart as Tenth Section" of *The Classic of Tea* recorded chart of tea cooking and twelve tea sets of all previous dynasties; the appendix part included tea-related laws and regulations since the Dezong Reign of the Tang Dynasty. Half-fermentative Oolong tea appeared accidentally, it was just one step away from the complete fermentative black tea of which the earliest type—Souchong black tea arose in Chong'an, Fujian Province in the 17th century, then the Congou black tea was loved gradually by people in every part of China. Longjing tea from Zhejiang Province gained a widespread fame after Emperor Qianlong tasted it when he made an inspection trip in south China.

褐色，也称为黑茶。现在流行的普洱茶仅仅是过去云南普洱茶的一种，为紧压茶或黑茶类。

散茶的流行和黑茶的出现在茶道审美上，某种程度地颠覆了中国传统茶道对绿茶对真香真味的追求，但其中的茶道精神和茶道思想并没有后退和改变。不论杯中的茶汤如何变幻，但茶人追求俭德清雅、崇尚奉献的懿德节操没有褪色。这是中国茶道文化能够与时俱进的魅力所在。因为中国茶道文化是中国优秀传统文化的有机组成部分，必将随着中国文化的进步，不断焕发新的生机。

因此古代饮茶有五种：（1）瓶缶煮饮，加各种调料；（2）瓶缶沃淹，加各种调料；（3）缶中烹煮只加少量盐而已；（4）点茶，用执壶冲注盏中筛罗好的茶粉；（5）散茶冲泡，既可以用碗，又可以用瓶、用壶。陆羽虽然提倡宫廷重要场合都应采用缶中烹煮，但其他饮茶法并没有在社会上消失，反而一直延续。例如湖南、江西、福建的擂茶，在陆羽时代就已经存在很长时间。舂，是擂茶必不可少的过程。我们之所以把瓶中煮饮和瓶中煮淹沃分开来，是因为茶瓶在中华茶道中一直使用，既可以烧水、煮茶，又可以沃茶（宋人又称为瀹茶）。对于

Dark tea, a kind of pile fermented and tightly compressed tea came into being since the Ming and Qing dynasties in order to adapt to the tea export. Pu'er tea popular now is just one sort of tightly compressed teas from Yunnan in the past.

The popularity of loose tea and the emergence of black tea subvert Chinese people's tradition of persuing real fragrant taste of green tea in the aesthetic aspect while remaining the core connotation of Chinese tea culture. No matter how the color of tea in the cup changes, the moral principle of thrifty, rhythmic tranquility, elegant, and dedication cherished by Chinese tea man is still there. This is the charm of Chinese tea culture of keeping pace with the times. Tea culture is an organic part of excellent Chinese traditional culture and will bound to glow constantly new vitality with the improvement of Chinese culture.

There were five kinds of tea-making and tea-drinking method in ancient China. 1. Cook tea in pottery vessels and add various sorts of spices. 2. Make tea in pottery vessels with boiling water and addvarious sorts of spices. 3. Cook tea in pottery vessels but add just a small amount of salt. 4. Make tea by means of pouring boiling water into tea dust. 5. Make tea by pouring boiling water into loose tea leaves. The other tea-making methods still existed in the society for instance Leicha (ground tea) which had been popular in Hunan, Jiangxi, and Fujian provinces for a long time although. Lu Yu advocated that tea should be cooked in pottery Fou-vessel in important royal court occasions. Tea bottle has been used in China for long, so we talk about cooking tea in bottle and making tea in bottle in here separately. People of the Song Dynasty called the method of making tea with boiling water "Yue", we will discuss the meaning of tea-making-related terms "Wo", "Yang", and "Yan", used in the Song Dynasty in the concerned chapter. We can

"沃""炀"或"痷"茶，我们在宋代茶道中加以论述。依据《茶经》，我们简单地勾勒出"烹茶道"基本程序。

但在这里必须指出，汉景帝阳陵（前141年封闭）出土的茶叶，被吉尼斯世界纪录确定为世界最早的茶叶。我认为这是茶开始进入饮用阶段的标志。

draw a simple outline of the basic tea-cooking procedures according to *The Classic of Tea*.

Here need to point out, the tea (sealed in 141BC) unearthed from Yang mausoleum of emperor Jingdi of the Han Dynasty has been confirmed as the earliest tea in the world by Guinness World Records. I think the tea marks thebeginning of drinking tea.

图 001

汉景帝（刘启，前 188 年—前 141 年）阳陵出土茶叶样品陆羽为了研究茶叶，遍览各种历史资料，从汉代王褒的《僮约》到《南北史》都进行了阅读。将历代茶事历史归纳进《茶经》第七部分的《七之事》，茶事人物从炎帝神农氏到南齐世祖武皇帝到陶弘景的茶事活动都进行了简要勾勒。

从第七部分我们看到，茶已经从普通饮品进入到中国文化领域，煎茶品茗成为诗词歌咏诵的题材，齐武帝以茶示俭，晋陆纳、桓温以茶示廉。

《茶经》茶道示意图

1. 启封

2. 赏茶饼

3. 炙茶

4. 碎茶饼

5. 碾茶

6. 筛罗

第一章

陆羽湖州茶文化圈
与茶道文化的诞生

Chapter One
Tea Culture Circle in Huzhou
and Birth of Tea
Ceremony Culture

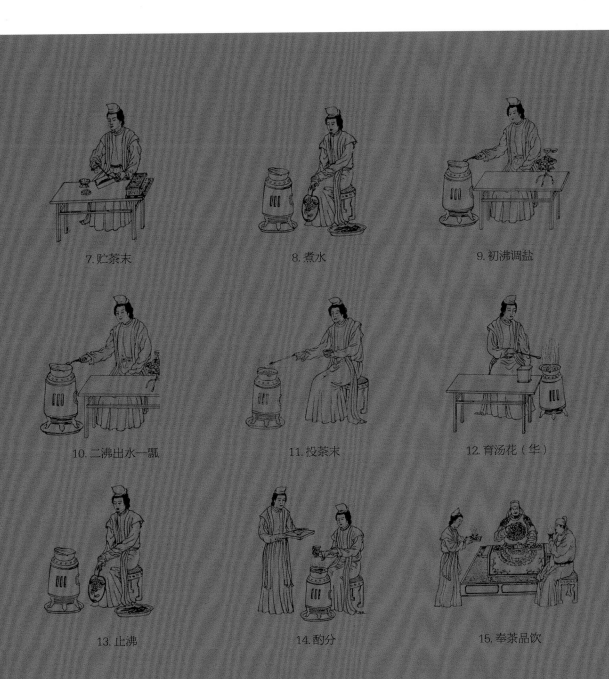

7. 贮茶末

8. 煮水

9. 初沸调盐

10. 二沸出水一瓢

11. 投茶末

12. 育汤花（华）

13. 止沸

14. 酌分

15. 奉茶品饮

浙江湖州是南朝至隋唐前期江南地区的经济文化重镇。所谓"苏湖熟，天下足"，这里的"苏湖"指的就是苏州和湖州。特别在陆羽定居、颜真卿刺湖和当地僧皎然、灵一、灵澈及唐代大历诗人频繁唱和的大历—贞元时期，这里形成中国第一个茶人集团，在特定的历史条件下，使茶道文化日臻成熟，并迅速达到历史上第一个高峰。

一、《茶经》初稿上元初至大历前就已经流布于社会

除正史"两唐书"外，有关陆羽生平及《茶经》介绍，还散见于陆羽自撰《陆文学自传》及封演《封氏闻见记》、李肇《国史补》、皮日休《茶中杂咏序》等唐代文献之中。《新唐书》载：

陆羽字鸿渐……上元初更隐苕溪，自称桑苎翁。……久之，诏拜羽太子文学，徙太常寺太祝，不就职，贞元末，卒。羽嗜茶，著经三篇，言茶之源、之法、之具尤备，天下益知饮茶矣。时鬻茶者至陶羽形置炀突间，祀为茶

Huzhou in Zhejiang Province was an important economic and cultural center in the Jiangnan Region from the Southern Dynasties to the Sui Dynasty and the early stage of Tang Dynasty. The so-called "A good harvest in Su and Hu can support the whole China" refered to Suzhou and Huzhou. China's first tea-man group took shape during Dali-Zhenyuan period when Lu Yu settled down in Huzhou, Yan Zhenqing was exiled to Huzhou, monks including Jiaoran, Lingyi, Lingche and other poets wrote responsorial poems about Huzhou. Then the tea culture gradually matured and rapidly reached the first peak in the history under the specific historical condition.

A. The First Draft of *The Classic of Tea* Had Been Spreading in the Society from the Early Shangyuan Reign to the Pre-Dali Reign

The introduction to Lu Yu's life and his book *The Classic of Tea* could also be found in literature of the Tang Dynasty such as *Autobiography of Lu Yu, What have been Seen and Heard by Feng Yan, Supplement for National History of Tang* by Li Zhao, *Prologue for Miscellaneous Poems on Tea Drinking* by Pi Rixiu in addition to the official history books *Old Book of Tang*and *New Book of Tang*. *New Book of Tang* has the following records:

Lu Yu whose style name was Hongjian... had a concealed life at Shaoxi in the early Shangyuan Reign and called himself elder peasant... he was given an official title by the emperor after a long time but he did not respond to that offer, died at the end of the Zhenyuan Reign. Lu Yu addicted to tea and wrote

神。有常伯熊者，因羽论复广著茶之功。御史大夫李季卿宣慰江南，次临淮，知伯熊善煮茶，召之，伯熊执器前，季卿为再举杯。至江南，又有荐羽者，召之，羽衣野服，挈具而入，季卿不为礼，羽愧之，更著《毁茶论》。其后尚茶成风，时回纥入朝，始驱马市茶。[1]

据此可知，上元辛丑岁（761），陆羽29岁时，已经写成三卷本的《茶经》。上元初，我们理解为上元元年，也就是760年。陆羽到了湖州就查阅资料，修改旧稿。

关于陆羽生平及《茶经》还应以与其关系密切的颜真卿、皎然等同时代人的交往为参考。

第一，很明确，颜真卿离开湖州是在五年任期末的大历十二年，也就是777年。[2]陆羽等一大批江南文化界精英参加了《韵海镜源》的编写。其间，陆羽对《茶经》进行充实，得以参阅丰富的文化典籍。

第二，皎然等与陆羽关系密切者。皎然，俗姓谢，字清昼，约生于唐玄宗开元八年（720），

three articles on the origin, method of classification and tea making utensils, so tea was known more by the public. The tea sellers at that time put the clay sculpture of Lu Yu in the shrine and worshiped him as the "Tea God". There was a man named Chang Boxiong practised tea cooking widely due to Lu Yu's theory on tea. The imperial censor Li Jiqing happened to know that Chang Boxiong was good at cooking tea while he inspected the region south of the Yangtze River, so he summoned Chang Boxiong and Chang Boxiong approached to him with tea utensils in hands, Li Jiqing raised his cup in order to show his respect to Chang Boxiong. Someone else also recommended Lu Yu to Li Jiqing while he was in this region, he summoned Lu Yu. Lu Yu visited him wearing casual clothes and taking tea utensils. Li Jiqing did not treat Lu Yu with courtesy as well, so Lu Yu felt ashamed, later wrote another book *On Ruining of Tea Ceremony*. Tea drinking became very popular thereafter, even the Uyghurs came to buy tea and carried it back with their horses.[1]

We know according to this that Lu Yu was 29 years old in 761, 2nd year of the Shangyuan Reign when he finished writing his 3-volume book *The Classic of Tea*.

Lu Yu's life and *The classic of tea* should be referred to the contacts between Yan Zhenqing, Jiaoran, etc. Who were close to him?

The first point is very clear that Yan Zhenqing left Huzhou in 777, 12th year of the Dali Reign, which was the end of his 5-year term of office.[2] A large number of culture elites from regions south of the

卒于德宗贞元末，享年 80 岁左右。皎然历来被公推为唐代诗僧之冠，如严羽所称："僧皎然之诗，在唐诸僧之上。"皎然也因此成了唐代大历诗人的重要代表之一。《全唐诗》收集的皎然之诗，除联句外共 474 首。其中至少有 300 余首是他在湖州写的或者是写湖州的，以湖州人文山水、民众疾苦、酬唱赠答的作品为最多。[3] 陆羽《陆文学自传》载：

……上元初，结庐于苕溪之滨，闭关对书，不杂非类，名僧高士，谈宴永日。……《茶经》三卷，占梦上中下三卷，并贮于褐布囊。上元辛丑岁子阳秋二十有九。[4]

第三，与陆羽同时代人的信息资料。进入正史记载的与陆羽有直接联系的是张志和。

张志和，字子同，婺州金华人。始名龟龄。……陆羽常问："孰为往来者？"对曰："太虚为室，明月为烛，与四海诸公共处，未尝少别也，何有往来？"[5]

另外，在萧颖士附传中，有其子萧存与颜真卿、陆羽的往来

Yangtze River including Lu Yu attended the compiling of the book *The Meaning of Ancient Chinese Echoism*. Lu Yu also enriched *The Classic of Tea* by using the opportunity provided only by such activities to read rich cultural classics while doing this work.

The second point is that Jiaoran remained a close relation with Lu Yu. Jiaoran, whose secular surname was Xie and style name Qingzhou, was born in 720, 8th year of the Kaiyuan Reign of the Tang Dynasty, and died at the age about 80 at the end of the Zhenyuan Reign of Emperor Dezong. Jiaoran had been regarded as the best monk poet of the Tang Dynasty just like Yan Yu said, "Poems written by monk Jiaoran are better than those by other monks of the Tang." Jiaoran thus became one of the most important representative poets of the Dali Reign of the Tang Dynasty. Among 474 Jiaoran's poems collected into *Full Collection of Tang Poems* at least 300 were about Huzhou or written in Huzhou especially those on Huzhou's landscape and social customs, common people's sufferings and responding friends' poems were in the largest amount.[3] *Autobiography of Lu Yu* has the following records:

"...Lu Yu built a cottage by the Shaoxi brook in the early Shangyuan Reign, stayed alone reading books, got along only with eminent monks and profound scholars, held feast and talked with friends all day long... *The Classic of Tea* contained three volumes... and was stored in brown cloth bag. Lu Yu was 29 years old in 761, 2nd year of the Shangyuan Reign."[4]

The third point is that the information about Zhang Zhihe who was Lu Yu's contemporary and had a direct

事迹：

萧颖士，开元二十三年举进士，对策第一。……建中初，（萧存）由殿中侍御史四迁比部郎中。[6]

从萧存的情形可以看出，他和陆羽相识、共事在建中（780—783）之前。

从李季卿宣慰江淮的时间看，早于大历时期（766—779）：

李适，京兆人，……子季卿，亦能文，……代宗立，还为京兆少尹，复授舍人，进吏部侍郎、河南江淮宣慰使。振拔幽滞，号振职。大历中，终右散常侍，遗命以布车一乘葬，赠礼部尚书。[7]

封演（生卒年不详），渤海蓨（今河北景县）人，天宝十五载（756）登进士第。贞元十六年（800）尚在世，约卒于贞元末。其《封氏闻见记》是距陆羽《茶经》时代最近的纪实作品：

楚人陆鸿渐为茶论，说茶之功效，并煎茶、炙茶之法。造茶具二十四事。以都统笼贮之。远近倾慕。好事者家藏一副。有常伯熊者，又因鸿渐之论广润色之。于是茶道大行。[8]

contact with Lu Yu which had been put into the official history.

Zhang Zhihe, whose style name was Zitong and initial name Guiling, was born into a family in Jinhua of Wuzhou... Lu Yu often asked, "Whom have you ever contacted with?" Zhang Zhihe answered, "I took the universe as my room, the moon as my candle and hosted all the people from my country, we had never departed, so why said with whom I have contacted?"[5]

In addition, the contact among Xiao Cun (son of Xiao Yingshi), Yan Zhenqing, and Lu Yu, was recorded in the attached part of *Biography of Xiao Yingshi*.

Xiao Yingshi, was chose as a successful candidate in the highest imperial examinations in 23rd year of the Kaiyuan Reign, took the first place in the argumentation on government affairs and Confucian classics of imperial examinations... was moved for four times from his post of imperial court censor to the post of tax and military supplies administrator in the early Jianzhong Reign of Emperor Dezong.[6]

It shows that Xiao Cun met Lu Yu and then they worked together before the Jianzhong Reign (780-783).

The imperial censor Li Jiqing inspected regions south of the Yangtze River before the Dali Reign (766-779).

Li Shi was from the capital... his son Li Jiqing was good at writing article... was appointed as a provincial governor, official department assistant minister, military and civilian affairs superintendent during the Emperor Daizong's Reign. He had not been promoted

其中"茶道"二字出现与皎然诗歌基本相差不远。封演生活在贞元末（805）以前。他的关于陆羽的记载应该是欧阳修《新唐书·陆羽传》的直接资料来源。常伯熊是陆羽茶道的第一个实践者和宣传者，"广"并"润色"陆羽之《茶经》茶道。

因此，我们认为，陆羽《茶经》是对开元时代茶事兴盛的理论概括，在上元二年（761）已经成书，并至迟在大历（766—779）以前在江南民间形成影响。

另外，独孤及（字至之，河南洛阳人）天宝末以道举高第，补华阴尉。代宗立，加检校司封郎中，徙常州刺史。卒年五十三，谥曰宪。其《慧山寺新泉记》不仅记述了陆羽对惠山泉闻名天下的推广作用，还明确表示"予饮其泉而悦之，乃志美于石"：

> 此寺居吴西神山之足……予饮其泉而悦之，乃志美于石。[9]

独孤及生卒年大约在725—777，其《慧山寺新泉记》记载了陆羽评鉴全国水泉的事实，但没有提及《茶经》。这有三种可能：（1）《茶经》已经如雷贯耳，

to a higher post although he was outstanding, was appointed as a government decree executant during the Dali Reign and given Minister of Rituals as posthumous title.[7]

Feng Yan (?-the end of the Zhenyuan Reign) was from Tiao (today's Jing County in Hebei) of Bohai and chose as a successful candidate in the highest imperial examinations in 756 died about at the end of the Zhenyuan Reign. His book *What Have Been Seen and Heard by Feng Yan* was an on-the-spot report work which was close to the times of *The Classic of Tea*. Lu Yu wrote the book discussing tea-related things including cooking tea, brewing tea, tea sets which amounted to 24 altogether and storage of tea wares. Those who loved tea tried all they could to have one tea set as personal collection. Chang Boxiong modified Lu Yu's manuscript thus making the tea ceremony very popular.[8]

These two words "tea ceremony" appeared almost at the same time when Jiaoran wrote tea-related poems. Feng Yan lived before the end of the Zhenyuan Reign (805), so his record about Lu Yu should be the direct data source for Ouyang Xiu's manuscript "Biography of Lu Yu" in *New Book of Tang*. Chang Boxiong was the first practitioner and propagator of tea ceremony established by Lu Yu also the person who retouched *The Classic of Tea*.

Therefore, *The Classic of Tea* was a theoretical summary on prosperity of tea activity during the Kaiyuan Reign of the Tang Dynasty and finished writing in 2nd year of Shangyuan Reign (761) thus impacting social life in regions south of Yangtze River no later than the Dali

不必赘述；（2）或此记以惠山泉为主，不及《茶经》，乃文意使然；（3）《茶经》这时（777年以前）影响有限。

二、《茶经》在湖州增补修订过

首先，从《茶经》的结构体例来看，其上、中、下三卷，从《一之源》到《六之饮》，已经定型。这可能是大历之前用挂图形式已经在茶肆悬挂坊间传抄的内容。《七之事》的主要内容就是搜集整理历史上的茶事历史。完成这样宏大的历史探索，是任何一个民间人士所无法想象的。只有借颜真卿编纂《韵海镜源》这个机会，方可做到。

其次，从现存《茶经》看，在上、中、下三卷中，下卷明显比重超大。包含五、六、七、八、九、十等六个小节。我们推测，这是因为七、八、九、十这四个小节的相当一部分是在湖州充实进来的。

最后，陆羽之所以将《八之出》，放在《七之事》之后，也是因为在 766 年以前，或者说在

Reign (766-779).

Moreover, Dugu Ji, who was from Luoyang of Henan Province and whose style name was Zhi Zhi, passed the imperial examinations at the provincial level during the Tianbao Reign of Emperor Xuanzong and was given a post of military commandant. Then he worked in the Ministry of Official Personnel Affairs and as a provincial governor of Changzhou in the Emperor Daizong's Reign. Dugu Ji died at age of 53 and was called Xian as his posthumous title. His article "New Springs at Huishan Temple" recorded not only Lu Yu's contribution in popularizing the fame of springs at Huishan Mountain but also expressed clearly that "I felt happy when drinking the spring water of Huishan Mountain and watching beautiful stones".[9]

Dugu Ji (725-777) recorded the evaluation onspring water all over Chinain his article "New Springs at Huishan Temple", but didn't mention Lu Yu's book *The Classic of Tea*. There are three possibilities for that: *The Classic of Tea* had been very popular all over the country, so there was no need to mention it; Dugu Ji gave priority in his article to the spring at Huishan mountain leaving no place for other contents; the influence of *The Classic of Tea* was limited (before 777).

B. *The Classic of Tea* Was Once Supplemented and Revised in Huzhou

Firstly, the structure and style of the three volumes of *The Classic of Tea* containing 6 chapters has already been fixed. So this was the content which had been copied widely in form of wall chart in many tea shops and tea houses. The main content of the 7th chapter of

761 年以前，陆羽不一定遍尝所有茶类，更不可能走遍所有茶区，也不可能搜集到理想的足够齐全的存世书籍。

三、颜真卿编写《韵海镜源》时的湖州茶人集团

颜真卿《湖州乌程县杼山妙喜寺碑铭》为我们提供了有关编写《韵海镜源》的直接资料，因是颜真卿本人撰文，故有极高的信史价值：

大历七年（772），真卿蒙刺是邦。

《韵海镜源》，自秋徂春。编同贯鱼，学比成麟。幸托胜引，亟倍僧珍。庶斯见传，金石不泯。[10]

可以看出，在湖州先后参与编修者超过 43 人。

就时间而言，大历七年（772）夏天开始，来年春完成此事。

此碑铭另一个价值是关于袁高的记载。我们知道顾渚贡茶院有袁高建中二年（781）的题记刻文。《全唐文》有陆贽举荐袁高的奏章。

令狐峘的《光禄大夫太子

The Classic of Tea was the tea activity happened in the history. It was unimaginable for any citizen to make such grand historical exploration which could only be completed with the precious opportunity provided by the compilation of *The Meaning of the Chinese Ancient Echoism* organized by Yan Zhenqing.

Secondly, it was obvious that the third volume of *The Classic of Tea* containing the 5th–10th chapters was oversized than the other two volumes, so the 7th–10th chapters of *The Classic of Tea* might be supplemented in Huzhou.

Thirdly, the reason for Lu Yu's putting chapter 8 after chapter 7 was that Lu Yu had not tasted all types of tea, had not possibly travelled to all tea-producing regions and had not been possible to collect completely all tea-culture-related books preserved till then before 766 or 761.

C. The Tea-man Group in Huzhou at the Time When Yan Zhenqing Was Compiling *The Meaning of the Chinese Ancient Echoism*

The "Stone Stela Inscription in Miaoxi Temple at Mount Zhushan in Wucheng County of Huzhou" written and copied by Yan Zhenqing provided us a direct data about the compiling of *The Meaning of the Chinese Ancient Echoism*, it was a trustworthy historical account because it was written by Yan Zhenqing.

Yan Zhenqin was exiled to Huzhou in 7th Year of the Dali Reign (772).

The compiling of *The Meaning of the Chinese Ancient Echoism* continued from Autumn to Spring with great efforts, the related situation was copied onto the stone stela which would preserved forever, it

太师上柱国鲁郡开国公颜真卿墓志铭》：

> 大历末……公乃奏上所著《韵海镜源》，帝嘉之……[11]

殷亮《颜鲁公行状》云：

> ……公初在平原，未有兵革之日，著《韵海镜源》，成一家之作……及至湖州，以俸钱为纸笔之费。……撰定为三百六十卷。[12]

《韵海镜源》编辑工作结束后，在湖州府衙举行了盛大的庆典活动：

> 著书禅理化，奉上表诚信。探讨始河图，纷纶归海韵。[13]

《全唐诗》卷817—34皎然《奉和颜使君真卿修〈韵海〉毕，州中重宴》（节录）：

> 世学高南郡，身封盛鲁邦。九流宗韵海，七字揖文江。

另有《全唐诗》卷819皎然《奉陪颜使君修〈韵海〉毕，东溪泛舟饯诸文士》。

颜真卿刺湖之初，邀集士大夫与高僧大德参加编订《韵海镜源》，在当时的文化圈子来说确实是一个重要事件。但史学家更推崇他在安史之乱和李

would be cherished by the readers.[10]

Those who participated successively in the compiling of *The Meaning of the Chinese Ancient Echoism* in Huzhou were more than 43 people.

In terms of time, the compiling started in summer of 7th year of the Dali Reign and completed in the Spring of the next year.

Another value of this stone stela inscription is its record about Yuan Gao. As we know that Yuan Gao's engraved inscription of 2nd year of the Jianzhong Reign (781) had been preserved in the Guzhu royal tribute tea processing plant. *Full Collection of Tang Prose* has Lu Zhi's document for recommending Yuan Gao as its content.

"Grave Stone Inscription of Yan Zhenqing" written by Linghu Huan recorded:

At the end of Dali Reign... Yan Zhenqing dedicated *The Meaning of the Chinese Ancient Echoism* to the emperor. The emperor rewarded him...[11]

"Biographical Sketch of Yan Zhenqing" by Yin Liang contains the following content: "...Yan Zhenqing was originally in the Central Plain and engaged in the compiling of the book *The Meaning of the Chinese Ancient Echoism* when there was no war... finished 360 volumes while in Huzhou."[12]

A grand celebration ceremony was held after the completion of the compiling of *The Meaning of the Chinese Ancient Echoism*. We compiled this book discussing origin of ancient Chinese echoism at the same time showing our love for ancient culture of our native country.[13]

The following content is the excerpt from Jiaoran's

希烈叛乱中所表现出来忠君、爱国情操以及以身践行儒家价值观的伟岸形象。因此，在"两唐书"人物传记中，只字未提编修《韵海镜源》。但我们看到正是由于颜真卿在士大人阶层包括高僧大德中享有崇高威信，才使得编纂文化典籍的工作，应者云从，群英荟萃。

四、品茗论道的茶文化圈

因编撰《韵海镜源》在湖州形成的这个文化圈子并没有因编纂工作完成而各自离去。应该说至少持续到颜真卿离开的大历末，即大历十三年（778）以后，由上述段亮文可知，颜真卿在此撰有《吴兴集》十卷。我们从中可知皎然、陆羽、皇甫兄弟和张志和及湖州寺院道观里的高僧、道士每有唱和、茶宴进行。

1. 陆羽的圈子

这个茶人圈子自然以颜真卿、陆羽和皎然为核心。如果说颜真卿因为在"安史之乱"中的杰出表现而获得全国性的影响，那么陆羽在茶文化界的

article "On Celebration of the Completion of the Compiling of the book *The Meaning of the Chinese Ancient Echoism*" in *Full Collection of Tang Poems*. Yan Zhenqing, a man of learning, did all he could to compile *The Meaning of the Chinese Ancient Echoism* to find by hard and thorough search the source of ancient Chinese echoism.

The following excerpt from Jiaoran's article "Farewell Dinner on Boat for All Scholars Who Participated the Compiling of *The Meaning of the Chinese Ancient Echoism* When the Work Was Completed" in volume 819 of *Full Collection of Tang Poemsrecords*. In the early stage of Yan Zhenqing's exiling to Huzhou, he invited scholars, officials, and eminent monks to compile *The Meaning of the Chinese Ancient Echoism* which arose a stir in the cultural circle at that time. But historians praised highly that Yan Zhenqing displayed loyal, patriotism, practicing Confucian values in the An Lushan's Rebellion and Li Xilie's Revolt. Although no single word on the compiling and editing of *The Meaning of the Chinese Ancient Echoism* could be found in the biography section of two books— *Old Book of Tang and New Book of Tang*. It is for sure that the compiling work could only be conducted smoothly and successfully by all those participants who admired Yan Zhenqing as a man of dignity, loyalty, and patriotism.

D. View of Tea Culture Circle on Tea Tasting and Tea Ceremony

The specific cultural circle formed by compiling of *The Meaning of the Chinese Ancient Echoism* did not vanished when the work completed at least until the end of Dali Reignwhen Yan Zhenqing left Huzhou,

影响则也是如日中天。而颜真卿对陆羽的才学修养和茶学研究自然也很认可。

颜真卿《项王碑阴述》：

西楚霸王当秦之末，与叔梁避仇于吴，盖今之湖州也。……其神灵事迹，具见竟陵子陆羽所载《图经》。[14]

除独孤及、封演之外，与陆羽有直接接触及诗歌唱和的还有顾况、戴叔伦等。陆羽与他们的交往唱和我们在后边还有涉及。

2. 皎然的圈子

《宋高僧传》载，皎然生前与韦应物、卢幼平、吴季德、李萼、皇甫曾、梁肃、崔子向、薛逢、吕渭、杨逵等过从甚密，大家在一起或吟诗，或论道，交谊颇深。[15]

与皎然、陆羽直接交往者，有名有姓者60余人。其中颜真卿、张志和、袁高、梁肃、皇甫冉兄弟正史《新唐书》有传。而顾况、戴叔伦的诗歌表明，他们也与陆羽、皎然多有往还。

《宋高僧传》卷十五《唐余杭宜丰寺灵一传》载："与天台

and we can know from "Biographical Sketch of Yan Zhenqing" by Yin Liang that Yan Zhenqing had compiled what he wrote in Huzhou into a 10-volume collection called *Huzhou Collection* and published it in which the tea party, writing poems, replying each other in poems using the same rhyme sequence among Jiaoran, Lu Yu, Huangfu Ran, Huangfu Zeng, Zhang Zhi, and some eminent monks from Buddhist & Taoist temples were recorded.

1. Lu Yu's Circle

Yan Zhenqing, Lu Yu and Jiaoran were naturally the core figures of this tea-man circle. If we say that Yan Zhenqing acquired national fame due to his outstanding performance in the An Lushan's Rebellion, then Lu Yu was a very influential figure in circle of tea culture. And that Yan Zhenqing also naturally recognized Lu Yu's talent, learning, and research on tea culture.

"Engraved Inscription on Back of Stone Stela of Xiang Yu" by Yan Zhenqing says: Xiang Yu and his uncle took refuge in Huzhou at the end of the Qin Dynasty... What he did could be found in *Picture Story of Huzhou* compiled by Lu Yu.[14]

Besides of Dugu Ji and Feng Yan, those who once had a direct contact with Lu Yu or replied each other in poems using the same rhyme sequence were Gu Kuang, Dai Shulun, etc. Lu Yu's contacted with them will be mentioned in the latter part of this book.

2. Jiaoran's Circle

Jiao Ran was born about in 8th year of the Kaiyuan Reign of Xuanzong Emperor (720) of the Tang

道士潘志清、襄阳朱放、南阳张继、安定皇甫曾、范阳张南史、吴郡陆迅、东海徐嶷、竟陵陆鸿渐为尘外之友。"[16]

灵澈（？—816）是一位律僧，曾著《律宗引源》二十一卷。吕温《张荆州画赞并序》云："曹溪沙门灵澈，岁脱离世务，而犹好正直。"[17]《宋高僧传》卷十五《灵澈传》数次提到灵澈与秘书郎严维、随州刘使君长卿等人唱和之事。戴叔伦《与友人过山寺》盛赞灵澈可与陶渊明相提并论，评价极高。……灵澈在大历诗僧中年辈较晚，卒于元和十一年（816），主要活跃于大历后期至贞元、元和中，其影响之大，还在灵一之上。[18]在当时，与皎然齐名的诗僧还有道标。

这个圈子在政治家之间也广有声势：相国李吉甫、司空严绶、右仆射韩皋、礼部侍郎吕渭、滑亳节制卢群、襄阳节制孟简、同州刺史李敷、凤翔尹孙（玉寿）、浙东廉事贾全、中书舍人白居易、随州刺史刘长卿、户部侍郎丘丹、外郎裴枢、秘阁严维、小谏朱放、越廉问薛戎、夕拜卢元辅、常州

Dynasty, died over the age of 70 between 9th and 14th year of the Henyuan Reign of Dezong Emperor. *Biographies of Eminent Monks of the Song Dynasty* recorded that Jiaoran had a very close relationship with Wei Yingwu, Lu Youping, Wu Jide, Li E, Huangfu Zeng, Liang Su, Cui Zixiang, Xue Feng, Lv Wei, Yang Kui, etc., reciting poems or discussing tea culture while they were together.[15]

There were over 60 persons who maintained an intimate relation with Lu Yu and Jiaoran among whom Yan Zhenqing, Zhang Zhihe, Yuan Gao, Liang Su, Huangfu Ran, and Huangfu Zeng have been included into the biographies of *New Book of Tang*. Gu Kuang and Dai Shulun recorded that they had contacted with Lu Yu and Jiaoran in their poems.

"Biography of Lingyi" in volume 15 of *Biographies of Eminent Monks of the Song Dynasty* has the following content, "Made real friends with Taoist monk Pan Zhiqing, Zhu Fang from Xiangyang, Zhang Ji from Nanyang, Huangfu Zeng from Anding, Zhang Nashi from Fanyang, Lu Xun from Wujun, Xu Yi from Donghai, and Lu Yu from Jingling."[16]

Lingche (?-816), a commandments-fast-hold monk, wrote a 21-volume book *The Source of the Vinaya School of Buddhism*. Lv Wen said in "On Appreciation of Paintings by Zhang Jingzhou and Its Preface" that: "Lingche, a man of integrity, once broke away fromworldly matters."[17] "Biography of Lingche" in volume 15 of *Biographies of Eminent Monks of the Song Dynasty* mentioned several times that Lingche, Yan Wei and Liu Changqing from Suizhou replied each other in poems using the same rhyme sequence.

释元浩、润州释南容、金华释乾辅、吴门释光严、上都释智崇等，与陆羽、颜真卿和皎然三人"心交尘外，分契林中"[19]。他们三人共同在世的时间也就在805年之前。这个圈子几乎涵盖了大历一贞元时代的相当一部分人，地域上横跨东西南北。

大历元年（766）常州义兴始贡茶，而大历五年（770），浙江长兴建贡茶院。贡茶成为一种制度。而皇帝将贡茶中的一些作为君臣和谐的媒介赐予大臣，《全唐文》中有数份《谢赐茶表》，以韩翃的较早。茶区友人在新茶上市时每每赶早寄茶给北方师友，有的同时附带信函。茶，成为友谊与信任的象征，成为诗人文友咏诵的高洁之物。许多茶诗名篇都是作者品尝到友人所寄新茶后的即兴之作，如中晚唐白居易的几首名作。

这些面对面的唱和与相互间的书信往还，将茶事放在突出地位，这也是历史上从未有过的文化盛况。这无疑是中国茶道文化经历开、天时代的普及与兴盛后，在思想内涵上日

Dai Shulun praised highly of Lingche who could be compared to Tao Yuanming in *Talking about Poetry with Friend in Mount Temple*. Lingche had been active from late period of the Dali Reign to middle period of the Zhenyuan and Yuanhe reigns, died in 11th year of the Yuanhe Reign (816). His influence was greater than that of Lingyi.[18] Daobiao, another monk poet was equally famous as Jiaoran at the time.

This circle exerted its momentum among politicians such as prime minister Li Jifu, great minister of Public Works Yan Shou, prime minister Han Gao, assistant minister of Ministry of Rites Lv Wei, commander of Huabo Lu Qun, commander of Xiangyang Meng Jian, prefectural governor of Tongzhou Li Fu, prefecture magistrate of Fengxiang Sun Yushou, Jia Quan from Zhedong, secretary Bai Juyi, prefectural governor of SuizhouLiu Changqing, deputy chief of the Ministry of Revenue Qiu Dan, assistant minister Pei Shu, royal bibliotheca librarian Yan Wei, provincial censor Zhu Fang, Xue Rong, imperial edicts communicator Lu Yuanfu, Shi Yuanhao from Changzhou, Shi Nanrong from Runzhou, Shi Qianfu from Jinhua, Shi Guangyan from Wumen and Shi Zhichong from Shangdu. These politicians maintained a very intimate relation with Lu Yu, Yan Zhenqing and Jiaoran.[19] Lu Yu, Yan Zhenqing and Jiaoran lived before 805, and those who had relation with this circle covered a certain proportion of renowned scholars from the Dali Reign to the Zhenyuan Reign across the country.

Yixing of Changzhou became the royal tea tribute producer in 1st year of the Dali Reign (766) while the institution for supervising the producing of royal tea tribute was established

臻成熟的文化土壤。

五、从诗歌看皎然对唐代茶道文化形成的贡献

从茶诗内容和对茶道文化的体悟来看，皎然的茶诗达到历史的新高度。

1. 初盛唐茶诗的回顾与评价

孟浩然（689—740）的《清明即事》可能是已知唐代最早的茶诗：

帝里重清明，人心自愁思。
车声上路合，柳色东城翠。

花落草齐生，莺飞蝶双戏。
空堂坐相忆，酌茗聊代醉。[20]

孟浩然的故乡是襄阳，陆羽眼中的中国第一茶区，也是风景优美适宜隐居的地方。他本人喜欢自然放达的隐士生活。[21] 他的这首诗，主题是长安清明节的见闻感受，充满哀愁、伤感、思忆，以茶代酒也与此情调极为贴合。而茶事在整个诗文中，只是点缀而已。《全唐诗》卷178李白《答族侄僧中孚赠玉泉仙人掌茶》《陪族叔当涂宰游化城寺升公清风亭》也提到茶："茗酌待

in Changxing of Zhejiang province in 5th year of the Dali Reign (770), and providing royal tea tribute gradually turned into a kind of system. Emperors usually bestowed ministers royal tea tribute as a medium of harmonious relation with them. Several official documents among which that written by Han Hong belonged to the early stage in *Full Collection of Tang Proses* could be regarded as the evidences of such medium. Those who lived in the tea producing area often mailed appeared-on-the-market fresh tea to their teachers and friends in the north, sometimes attached letter. Tea had become a symbol of friendship and trust, a noble and pure thing for which poets wrote and recited poems. Many famous tea poems were improvisational pieces of work after tasting fresh tea sent by good friends such as a few poems written by Bai Juyi in the middle and late Tang Dynasty.

The face-to-face poem writing and letter mailing put tea activities in the prominent position, this was spectacular cultural event which had not ever been seen in the history. This was undoubtedly the cultural soil for Chinese tea culture being mature in terms of its ideological content after experiencing popularization and thriving.

E. Jiaoran's Contribution to the Formation of Tea Culture of the Tang Dynasty in Terms of Poetry

Tea poems by Jiaoran reached a new height in history from the view of the tea poetry content and the awareness of tea culture.

1. Review and Evaluation of Tea Poetry in Early and High Tang Dynasty

幽客，珍盘荐凋梅。飞文何洒落，万象为之摧。季父拥鸣琴，德声布云雷。"

不言其诗歌艺术性，就茶事来说，李白只是停留在茶事茶叶这一茶文化的浅层阶段。初盛唐茶诗寥落且较为浅显。

陈兼的《代茶馀序略》则告诉人们，开、天时期，人们对饮茶所造成的致病之理已经有了客观的认识，而这些认识是长期观察、实践所得：

释滞销壅，一日之利暂佳；瘠气侵精，终身之累斯大。获益则归功茶力，贻患则不为茶灾，岂非福近易积知、祸远难见。[22]

2. 大历时代湖州之外的茶诗逐渐增多

大历十才子中，以钱起和韩翃的茶诗最为著名。

钱起（710—782），字仲文，吴兴（今浙江湖州）人，与郎士元、司空曙、李益、李端、卢纶、李嘉祐等称大历十才子。

钱起《过长孙宅与朗上人茶会》，可能是作者在北方时期的作品，如果是长安，这就是唐人

"My Feelings in the Tomb Sweeping Day" by Meng Haoran might be the earliest tea poem of the Tang Dynasty ever known. People in the capital city attached great importance to the Tomb-sweeping Day, everyone was in gloomy mood. The sound of wheels could be heard on the road, willow leaf made the eastern city green, flower fell while grass grew, warbler flew and butterfly played in pair, they sat in empty hall recalling past days when drinking tea also wishing to be drunk.[20]

Meng Haoran's hometown is Xiangyang, the No.1 tea producing area of China in the eyes of Lu Yu also a place with beautiful scenery which is suitable for seclusion.[21] Meng Haoran liked the natural and unrestrained hermit life, his poem "My Feelings in the Tomb Sweeping Day" mainly expressed what he saw, heard and felt in the Tomb Sweeping Day in Chang'an, his feeling filled with sorrow, sadness, recalling, his drinking tea instead of alcohol being fitting for the emotional appeal in a such day. The tea activity serves just as an ornament in this poem. The poems "On Cactus Tea from Mount Yuquan" and "Sightseeing Breeze Pavilion of Huacheng Temple" by Li Bai from volume 178 of *Full Collection of Tang Poems* have tea-related content: "Entertaining my uncle quietly with tea, preserving withered plum in my precious dish; my uncle's gifted aptitude for writing eclipsing the colors of all things on earth, the music he played sounding like thunder in the clouds."

We are not going to discuss the artistic quality of Li Bai's poems, just talk about the tea-related content of his poems because Li Bai wrote tea activity on a

在长安写的第一首以茶会为主题的诗歌。

钱起《与赵莒茶宴》则是较早的茶宴诗。仲夏午后，诗人与赵莒等举行茶宴："竹下忘言对紫茶，全胜羽客醉流霞。"《全唐诗》卷239钱起《过张成侍御宅》："杯里紫茶香代酒，琴中绿水静留宾。"[23]。

刘长卿（约726—约786）也有茶诗留存。其《惠福寺与陈留诸官茶会（得西字）》明确告诉我们：事件是茶会，地点在陈留（今开封地区境内）的惠福寺。这是在北方出现较早的有关茶会的记载。

3. 大历诗人与江南—湖州茶文化圈的因缘来往

许多大历诗人与陆羽、皎然、灵一、灵澈和道标有诗赋往还，或者直接与他们唱和，或来江南来湖州公干、游历，受到湖州茶文化圈的耳濡目染。

湛然（711—782），唐代天台宗高僧，俗姓戚，常州晋陵荆溪（今江苏宜兴县）人。《宋高僧传》云："天宝末，大历

lower level of tea culture. It shows that the tea-related poems of early and high Tang Dynasty are limited in quantity and relatively shallow in connotation.

Chen Jian said in his article "As a Dinner Talk Preface" that people have gained the objective knowledge on pathology caused by drinking tea through long-term observation and practice during the Kaiyuan and Tianbao reigns. "Dispelling the short-term stagnation to be feeling good for one day, however the loss of strength will turn out to be a long-term weariness. One can benefit from drinking tea, sufferings left behind should not be attributed to tea. Would not it be easy for one to know the blessing which is near while hard to foresee the disaster which is far?"[22]

2. Tea-related Poems Increased Outside of Huzhou in the Dali Reign

Tea-related poems written by Qian Qi and Han Hong are the most famous ones among those by ten gifted scholars in the Dali Reign.

Qian Qi (710-782) with style name Zhongwen is from Wuxing(today's Huzhou of Zhejiang Province) and called ten gifted youth together with Lang Shiyuan, Sikong Shu, Li Yi, Li Duan, Lu Lun, Li Jiayou, etc.

The poem "Three-person Tea Party" might be written by Qian Qi while he stayed in the north, if it was Chang'an where he stayed, this poem should be the first tea-party poem composed by Tang people in the capital city.

The poem "Two-person Tea Feast" by Qian Qi belonged to earlier period work of its kind. It described a scene like this: Qian Qi and Zhao Ju held a

初。诏书连征。……建中三年二月五日示疾佛陇道场。……春秋七十二。法腊三十四。……其朝达得其道者唯梁肃学士，故擒鸿笔成绝妙之辞。"[24]

我们仅从诗歌题目就可以看到，湖州茶文化圈无疑具有全国性的影响。

六、湖州茶风

1. 陆羽茶文化实践

陆羽的茶文化实践包括到各地采茶制茶鉴定茶品质高下，同时也对烹茶之水进行品评。其足迹以湖州为中心，四处探查采制。

皇甫冉《送陆羽之越序》：

君仓穷孔释之名理，极歌诗之丽则。远墅孤坞，通舟必行，鱼梁钓矶，随意而往。夫越地称山水之乡，辕门当节钺之重。鲍侯知子爱子者，将解衣推食。岂徒尝镜水之鱼，宿耶溪之月而已？[25]

以颜真卿为盟首，以陆羽、皎然为核心的湖州茶文化圈是当时唐朝最活跃、人数最多的新文化圈子，是具有全国影响的茶人集团。

tea feast in a mid-summer afternoon, "Having a happy talk while drinking tea in bamboo forest, we both are drunk, so seems to see floating rosy clouds before our eyes." Tea-related description can also be found in the poem "Lingering around Zhang Chengshi's Home" by Qian Qi from volume 239 of *Full Collection of Tang Poems*, "Taking the fragrant tea in cup as alcohol, wonderful music quietly asking the guest to stay."[23]

Tea-related poems composed by Liu Changqing (about 726-about 786) have been preserved and handed down till today. His poem "Tea Party at Huifu Temple" tells us clearly that the tea party was held at Huifu Temple in Chenliu (today's Kaifeng). This is the tea-party record of early period in the north.

3. Karma and Contact between Dali-Reign Poets and Tea-Culture Circle of Huzhou in Regions South of the Yangtze River

Many Dali-Reign poets remained contact with Lu Yu, Jiaoran, Lingyi, Lingche and Daobiao by means of writing or face-to-face discussing poetry and prose, or had been influenced by what they saw, heard or experienced in Huzhou tea culture circle when they came to Huzhou on business or for travel.

Zhanran(711-782), a Buddhist monk of Tiantai-Sect in the Tang Dynasty whose worldly surname is Qi, is a native from Jingxi of Jinling in Changzhou (today's Yixing City, Jiangsu Province). *Biographies of Eminent Monks of the Song Dynasty* records, "The imperial edicts were issued continuously at the end of the Tiaobao Reign and the beginning of Dali Reign... Zhanran was ill on February 5th of 3rd year of

2. 湖州茶人的高雅生活

我们从颜真卿的一首诗说起。大历后期的某年某月某日，在杼山上，看到满眼绿色，桂花飘香，陆羽大生感慨，不由想起颜真卿，便随手折下一束带着青枝的桂花，寄送给做刺史的颜真卿。颜鲁公极为感动便写诗，以作答谢。君子之交，如此淡泊、清雅，又不乏知己互敬的深情。

颜真卿刺湖，无疑给这个江南小城带来一股清风，他们饮茶赋诗，谈古论今。

湖州茶人的生活高雅、宁静淡泊而又充满向上的文化情怀。

七、皎然诗所提倡的茶道思想的高度

如果说中国茶道文化崛起于湖州文化圈的话，那么陆羽以茶道文化实践考察为主，皎然则以茶道文化的设定和茶诗创作为特色，以他们为核心的茶文化圈共同把茶道文化推上新的阶段。

孟浩然、李白与皎然的茶诗有明显的差距，这是时代使然；

Jianzhong Reign... He was 72 years old... He believed that the only one who understood the essence of the doctrine was Liang Su who could compose excellent poetry and prose."[24]

We can see only from the topic of poetry that Huzhou tea culture circle had undoubtedly a nationwide influence.

F. Tea Drinking Custom in Huzhou

1. Lu Yu's Tea Culture Practice

Lu Yu's tea culture practice included picking tea leaves, manufacturing tea, making an appraisal of tea quality, tasting, and evaluating tea-cooking water. Lu Yu had set his foot in many places centered mainly in Huzhou.

Huangfu Ran wrote "Preface for Seeing Lu Yu off to Yuezhong" saying: "Lu Yu studied the doctrine of Confucianism and Buddhism, sought the beauty of poem. One can go to the distant cottage or lonely island if there are ships. Yuezhong is a famous place with wonderful landscape, and Junmen is a place of military importance. Mr. Bao knows you and loves your talent, he will take a good care of you. You should not just taste the fish from Jing lake and only appreciate the moon of Yexi!"[25]

The Huzhou tea culture circle with Yan Zhenqing as its leader, Lu Yu and Jiaoran as its core members was the largest and most active new culture circle having a nationwide influence in the Tang Dynasty.

2. Elegant life style of tea man in Huzhou

Let us talk about one of Yan Zhenqing's poems. One day in late Dali Reign, Lu Yu saw eyeful green and

同时代的刘长卿、钱起、耿湋及顾况、戴叔伦的茶诗与他也有距离，他们对茶事和茶道文化的理解没有皎然细腻而深刻。

皎然的因素不必多言，而身为唐朝大儒的颜真卿也是在家佛弟子[26]，这些写过茶诗，与高僧大德的诗歌往还，喜欢品茗的文人墨客，对禅有比较深刻的理解。这些历史事实对中国茶道文化的多元化产生积极影响。

湖州茶文化，对此后杭州刺史白居易的茶诗创作、张又新的《煎茶水记》的完成，对裴汶写成《茶述》都具有直接影响。

小结

《茶经》初稿在 761 年前已经在民间流传，按照《李季卿传》推算，在大历以前，即 766 年以前，《茶经》已经被常伯熊等推广演绎；《茶经》最终定稿时间在大历八年至十二年（773—777）之间。而《茶经》正式向全国发行，当在颜真卿大历末在刑部、吏部任职期间，最迟在 781 年杨炎专权后他被降为太子太师之前。

安史之乱之后的李希烈、刘

smelled osmanthus fragrance on Mount Mao, sighed with emotion and could not help thinking of Yan Zhenqing, then he randomly picked one osmanthus with tender green branch and sent it to Yan Zhenqing later. Yan Zhenqing was extremely touched and wrote a poem showing his gratitude. The friendship between gentlemen was so elegant full of deep feeling of great respect.

Yan Zhenqing undoubtedly brought a fresh wind to this tinny town of Jiangnan when he was exiled to Huzhou namely led a fashion of local residents to write poem and prose, drink tea and talk about the past and today.

The daily life of tea men in Huzhou was elegant, quiet, not seeking fame and wealth and full of cultural feelings of optimism.

G. The Height of the Thoughts on Tea Culture Advocated by Jiaoran with His Poems

If we say that China's tea culture rose from culture circle in Huzhou, then the tea culture circle centered on Lu Yu and Jiaoran pushed up Chinese tea culture to a new stage with their diligent effort such as Lu Yu's tea culture practice and related investigation and Jiaoran's giving definition to tea culture and writing tea-related poems.

There are obvious differences due to different eras among tea-related poems written by Meng Haoran, Li Bai, and Jiaoran; tea-related poems by contemporary poets including Liu Changqing, Qian Qi, Geng Wei, Gu Kuang, and Dai Shulun are not the same as that of Jiaoran because they could not understand the tea activity and tea culture as exquisitely and profoundly.

展叛乱对以扬州为中心的江南、以襄阳为中心的江南西道、以蔡州—许州为中心的淮西造成巨大破坏,加上扬州交通地位下降[27],浙西浙东经济文化地位在代宗、德宗朝显著上升,这是湖州成为贡茶基地的大背景。

陆羽《茶经》初有影响,颜真卿刺湖而群英荟萃,皎然长于思考和诗歌创作,三者历史性地有机结合促成了以他们为核心的中国第一个茶文化圈的形成,进而促进茶文化新阶段——茶道文化的日臻成熟,并形成第一次高峰。

茶道文化是对茶文化历史的提升与总结,既有茶事做基础以茶艺做依托,更有对世界对生命的态度和关照,是中国文化人在特殊背景下创造的新的文化形式。

【注释】

[1]《新唐书》卷一百九十六《隐逸·陆羽传》,中华书局,2011年,第5611—5612页。

[2] 郁贤皓:《唐刺史考全编》第三册,安徽大学出版社,2000年,第1935—1963页。

[3] 叶美芬:《诗僧皎然的生平及

Poets and literary men including Yan Zhenqing who was an at-home Buddhist disciple created tea-related poems,[26] wrote poems and prose as answers to those by eminent monks, liked drinking tea, understood profoundly the dhyana thus made a positive contribution in forming Chinese tea culture with diversified elements.

Huzhou tea culture exerted direct influence on the creation of tea-related poems by Bai Juyi, the writing of *Water for Tea Brewing* by Zhang Youxin and the completion of *On Making Tea* by Pei Wen.

Summary

The first draft of *The Classic of Tea* by Lu Yu had been spreading in the society before 761, popularizing by Chang Boxiong before the Dali Reign (766) calculated according to *Biography of Li Jiqing*, finalizing from 8th to 12th year of the Dali Reign (773-777) and issuing at the end of the Dali Reign when Yan Zhenqing held a post in Ministry of Penalty and Ministry of Official Personnel Affairs or before 781.

Huzhou became the first royal-tribute tea producer in the background of Li Xilie and Liu Zhan's Revolt after An Lushan's Rebellion causing tremendous destruction in Yangzhou-centered Regions South of the Yangtze River, Xiangyang-centered Jiangnan West Road and Caizhou-and-Xuzhou-centered Huaixi, Yangzhou's status as traffic hub declined.[27] The economic and cultural status of eastern Zhejiang and western Zhejiang went up in the Daizong and Dezong reigns.

The following three aspects— Lu Yu's *The Classic of Tea* had the preliminary effect; Yan Zhenqins was

湖诗的考究》,《湖州职业技术学院学报》2006年第2期。

[4]［唐］陆羽:《陆文学自传》,《全唐文》卷四百三十三,中华书局,1983年,第4420—4421页。

[5]《新唐书》卷一百九十六《列传一二一·隐逸》,第5608—5609页。

[6]《新唐书》卷二百零三《艺文下·萧颖士》,第5770页

[7]《新唐书》卷二百零二《艺文中·李适附李季卿传》,第5748页。

[8]［唐］封演:《封氏闻见记》,赵贞信校注,中华书局,2005年,第51页。

[9]［唐］独孤及:《慧山寺新泉记》,《全唐文》卷三百八十九,中华书局,1983年,第3950页。

[10]［唐］颜真卿:《湖州乌程县杼山妙喜寺碑铭》,《全唐文》卷三百三十九,第3435页。

[11]［唐］令狐峘:《光禄大夫太子太师上柱国鲁郡开国公颜真卿墓志铭》,《全唐文新编》第7册,吉林文史出版社,1999年,第4515页。

[12]［唐］殷亮:《颜鲁公行状》,《全唐文》卷五百一十四,第5223页。

[13]《全唐诗》卷八百一七《奉和颜使君真卿修<韵海>毕会诸文士东堂重校》,《全唐诗》卷八百一十七《皎然三》,中华书局,1980年,第9201页。

[14]［唐］颜真卿:《项王碑阴述》,《全唐文》卷三百三十八,第3432页。

[15]陈云琴:《诗僧皎然与茶文化》,《农业考古》2006年第2期。

[16]［宋］赞宁:《宋高僧传》卷十五《唐余杭宜丰寺灵一传》,中华书局,1987年,第360页。

[17]［唐］吕温:《张荆州画赞并

exiled to Huzhou; Jiaoran was good at thinking deeply and poetry creation— had promoted the forming of the first Chinese tea culture circle, accelerated the maturing of the tea culture at the same time, and then facilitated the coming of its first peak.

The tea culture is a kind of promotion and summary of that in the past dynasties based on tea-related ceremony and tea art meanwhile an expression of Chinese intellectuals' attitude towards the world and life also a culture form created by them under the special background.

【Notes】

[1]"Hermits", in *New Book of Tang*, Beijing: Zhonghua Book Company, vol.196, 2011, pp.5611-5612.

[2]Yu Xianhao, *Complete Collection of Research on Feudal Provincial Governor*, Hefei: Anhui University Press, vol.3, 2000, pp.1935-1963.

[3]Ye Meifen, "Monk Jiaoran's Life and Huzhou Poetry", *The Journal Huzhou Vocational and Technical College*, No.2, 2006.

[4]"Autobiography of Lu Yu", in *Full Collection of Tang Prose*, Beijing: Zhonghua Book Company, vol.433, 1983, pp.4420-4421.

[5]"Hermits", in *New Book of Tang*, Beijing: Zhonghua Book Company, vol.196, 2011, pp.5608-5609.

[6]"Xiao Yingshi", in *New Book of Tang*, Beijing: Zhonghua Book Company, vol.203, 2011, p.5770.

[7]"Biography of Li Shi and Li Jiqing" in *New Book of Tang*, Beijing: Zhonghua Book Company, vol.202, 2011, p.5748.

[8]Feng Yan, *What Have Been Seen and Heard by Feng Yan*, Beijing: Zhonghua Book Company, 2005, p.51.

[9]Dugu Ji, "New Springs at Huishan Temple", in *Full Collection of Tang Prose*, Beijing: Zhonghua Book Company, vol.389, 1983, p.3950.

序》，《全唐文》卷六百二十九，第6350页。

[18] 郜林涛：《大历诗僧灵一、灵澈述评》，《山西青年管理干部学院学报》2006年第3期。

[19]《宋高僧传》卷十五《唐杭州灵隐山道标传》，第374—375页。

[20]《全唐诗》卷一百五十九。有人将苏的"香名展骥初"，中的名视为"茗"，被视为最早茶诗。不确。"香名"是唐人的固定用法，指优良的名声，相同的用法还有"香位"，都有今天我们口语中的"吃香""红火"的含义。

[21] 叶嘉莹：《说初盛唐诗》，中华书局，2008年，第131页。

[22] 陈兼：《代茶馀序略》，《全唐文》卷三百七十三，第3792页。

[23] 绿应为渌，渌水为琴曲名。后来有白居易"琴里知闻唯渌水，茶中故旧是蒙山"比白居易诗早三十年。所谓"紫茶"指湖州紫笋茶，泛指上乘的茶。

[24]《宋高僧传》卷六《唐台州国清寺湛然传》，第117页。

[25]［唐］皇甫冉：《送陆羽之越序》，《全唐文》卷四百五十一，第4616页。

[26]《宋高僧传》卷二十六《唐湖州佛川寺慧明传》，第664页。

[27] 梁肃：《通爱敬陂水门记》，《全唐文》卷五百一十九，岁在戊辰，扬州牧杜公命新作西门，所以通水庸，致人利也。冬十有二月，土木之工告毕，从事徵其始，请刻石以为记云。……当开元以前，京江岸于扬子，海潮内于邗沟，过茱萸湾，北至邵伯堰，汤汤涣涣，无隘滞之患。其后江派南徙，波不及远，河流浸恶，日淤月填。若岁不雨，则鞠为泥涂，舟楫陆

[10]Yan Zhenqing, "The Stone Stela Inscription in Miaoxi Temple at Mount Zhushan in Wucheng County of Huzhou", *Full Collection of Tang Prose*, vol.339, p.3435.

[11]Linghu Huan, "Grave Stone Inscription of Yan Zhenqing", in *New Full Collection of Tang Prose*, Changchun: Jilin Literature Press, vol.7, 1999, p.4515.

[12]Yin Liang, "Biographical Sketch of Yan Zhenqing", in *Full Collection of Tang Prose*, vol.514, p.5223.

[13]*Full Collection of Tang Poems*, Beijing: Zhonghua Book Company, vol.817, 1980, p.9201.

[14] Yan Zhenqing, "Engraved Inscription on Back of Stone Stela of Xiang Yu", in *Full Collection of Tang Prose*, vol.338, p.3432.

[15]Chen Yunqin, "Monk Poet Jiaoran and Tea Culture", *Agricultural Archaeology*, No.2, 2006.

[16]Zan Ning, "Biography of Lingyi", in *Biographies of Eminent Monks of the Song Dynasty*, Beijing: Zhonghua Book Company, vol.15, 1987, p.360.

[17]Lv Wen, "On Appreciation of Paintings by Zhang Jingzhou and Its Preface", in *Full Collection of Tang Prose*, vol.629, p.6350.

[18]Gao Lintao, "Discussion and Evaluation on Monk PoetsLinyi and Linche of the Dali Reign", *Journal of Shanxi Youth Management Cadre College*, No.3, 2006.

[19]"Biography of Daobiao", in *Biographies of Eminent Monks of the Song Dynasty*, vol.15, pp.374−375.

[20]*Full Collection of Tang Poems*, Beijing: Zhonghua Book Company, vol.159, 1980.

[21]Ye Jiaying, *Talking about Poetry of Early and High Tang Dynasties*, Beijing: Zhonghua Book Company, 2008, p.131.

[22]Chen Jian, "As a Dinner Talk Preface", in *Full Collection of Tang Prose*, vol.373, p.3792.

[23]"Lu Shui" should be name of a piece of music and the socalled "Purple Tea" refers to the violet bamboo shoot tea from Huzhou which is also a high quality tea.

[24]"Biography of Zhanran", in *Biographies of Eminent Monks of the Song Dynasty*, vol.451, p.117.

[25]Huangfu Ran, "Preface for Seeing Lu Yu off to Yuezhong",

沈，困于牛车，积奥含败。人中其气，
为疾为瘵。长民者时兴人徒，以事开
凿。既费累钜万，或妨夺农功，殚财竭
力，随导随塞，人不宽息，物不滋殖，
百有余年矣。

in *Full Collection of Tang Prose*, vol.451, p.4616.

[26]"Biography of Huiming", in *Biographies of Eminent Monks of the Song Dynasty*, vol.26, p.664.

[27]Liang Su, "Building New Watergate", in *Full Collection of Tang Prose*, vol.519.

图 002

唐·阎立本·《萧翼赚兰亭图》·辽宁省博物馆藏品

　　故事以萧翼诱骗僧人辩才出借王羲之《兰亭序》真品为主题。位于画面左侧的备茶的茶师
及其书童所用茶器，无疑是两宋的点茶道所用茶盏。点茶形式也许在唐初就有，但绝不是流行
形式。茶道形式包括茶器搭配，往往与画中人物所处时代不相吻合，是中国茶道画的基本情形，
也是茶学爱好者观赏茶道画所应该特别注意的常识。这很可能是假托阎立本的宋代作品。

图 003

唐·周昉（713—741）·《调琴啜茗图》

这是宫廷仕女抚琴奉茶的情景。雍容华贵的宫廷妇女，在树间抚琴、品茗，显得雅静、幽闲。这是中国宫廷妇女茶道画的基本样式。画中人物虽略显丰腴，但宁静、自信、矜持的神态，还是相当明显。

图 004

唐·周昉·《纨扇侍女图卷》·北京故宫博物院藏品

　　虽然图中没有一点品茗的内容，但有别于我们理解宫廷妇女的日常生活形态，也许这与前边《调琴啜茗图》在构思和画面意境上是一致的。绣花、熨烫衣服、熏衣，都是在平淡、安静的气氛中进行，其实与品茗异曲同工。（摄影／孙跃）

周昉·《纨扇侍女图卷》（局部）

图 005

唐·孙位·《高逸图》·上海博物馆藏品

　　绢本设色，高 45.2 厘米，宽 168.7 厘米。

　　三足盘中的三个白色瓷杯的形状，与穿白衣黑靴侍者手中的杯子，明显不同，很可能

是最早的"茶酒间进"图，站立侍者端的可能是茶汤，杯子稍大！

孙位·《高逸图》（局部）

图 006

唐墓壁画·《端杯仕女图》·新疆博物馆藏品

　　带高足茶托与茶盏的搭配是茶饮的一般组合形成。西域是中国饮茶沿丝路向西传播的重要
地区。西到中亚，南到南亚、吐蕃、诃陵。这是新疆出土的唐代壁画。

图 007

唐·佚名·《宫乐图》·台北故宫博物院藏品

　　该图是留存不多的描绘关于宫廷妇女生活情境的画作。从丰腴略显臃肿和主仆等级不同的人物体积差别较大的特点看，我们认为这是中唐作品。这些侍女正在饮用的很可能是茶汤。从手中碗的大小来看，直径在12—14厘米之间。任何酒浆如米酒、葡萄酒都不可能使用这么大的碗。

图 008

五代·顾闳中（约910—980）·《韩熙载夜宴图》·北京故宫博物院藏品

　　虽然称为"夜宴图"，实际主要是品茶喝酒、吃水果与茶点。因为床几之上并没有
筷子和勺子，也可能是宴饮之后的娱乐活动。（摄影/胡锤）

顾闳中·《韩熙载夜宴图》（局部）

熙載風流清
為天官侍郎以
傾為時論所誚
管聲此圖

图 009

法门寺茶器（一）

①法门寺素面盘旋座银盐台——贮存盐花

②法门寺淡黄绿色琉璃茶盏茶托——饮茶

③法门寺《物帐碑》拓片

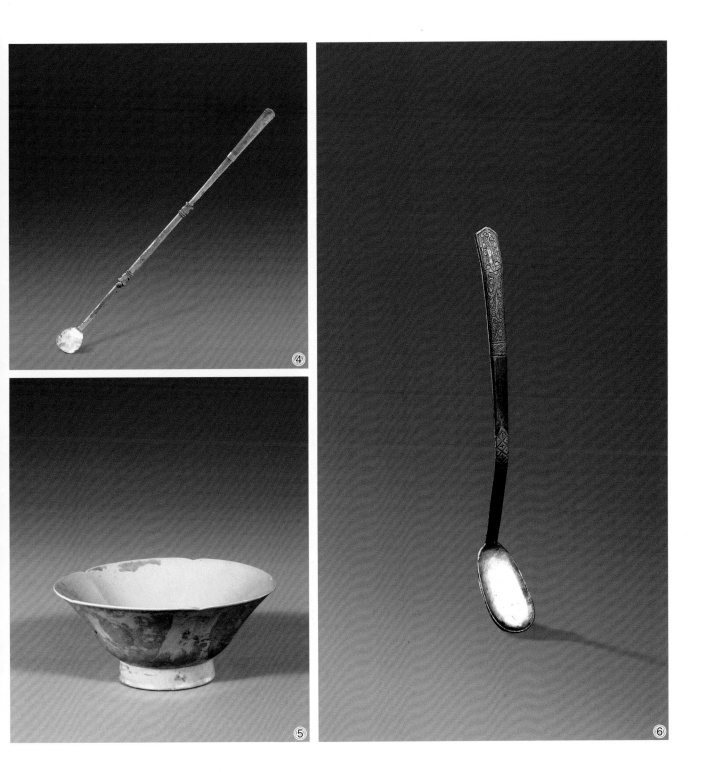

④法门寺鎏金团花长柄银勺——煮茶击拂

⑤法门寺五瓣葵口秘色瓷碗——点茶

⑥法门寺鎏金团花银则——取投茶末

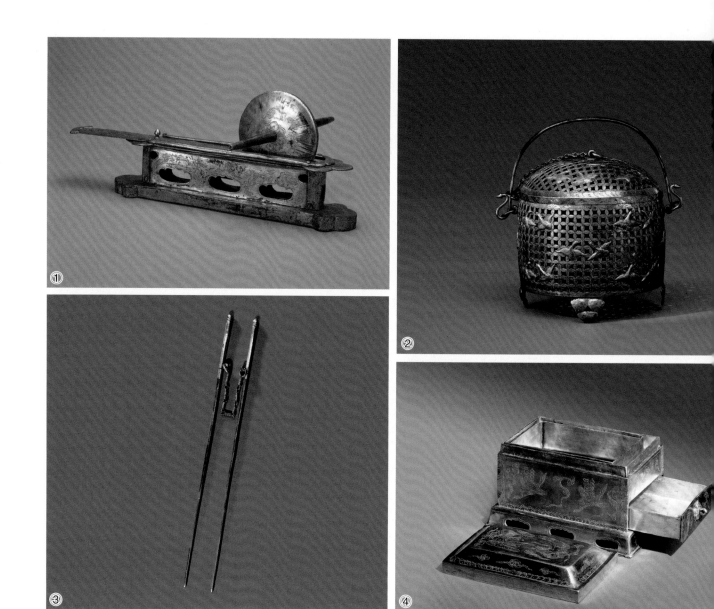

图010

法门寺茶器（二）

①法门寺鎏金团花银碢轴鎏金壶门座茶碾槽——碾茶

②法门寺鎏金飞鸿毬路纹银笼子——存贮茶饼

③法门寺鎏金系链银火箸——添加木炭

④法门寺鎏金仙人驾鹤纹壶门座银鉴罗——筛罗茶粉末

⑤法门寺鎏金团花银碢轴——碾茶

⑥法门寺金银丝结条笼子——贮茶饼

⑦法门寺鎏金摩羯纹三足架银盐台——贮盐花

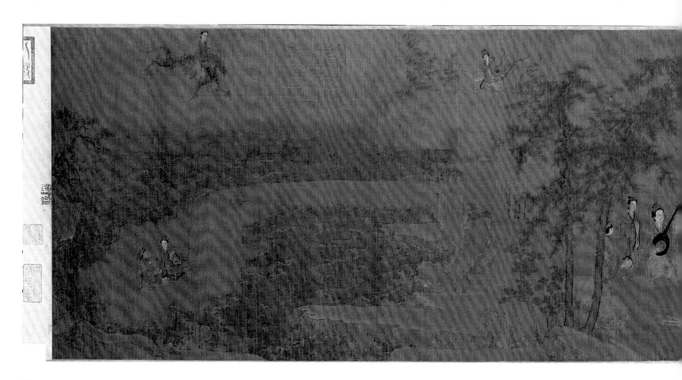

图 011

五代·阮郜·《阆苑女仙图卷》·北京故宫博物院藏品

　　绢本设色，高 42.7 厘米，宽 177.2 厘米。

　　阆苑，也称阆风之苑，在昆仑山之巅，为西王母所居之地。画家生活的时代正是饮茶之风炽盛之时，他不自觉地将饮茶想象为神仙的乐趣之一！此图与《宫乐图》都可视作宫廷妇女事茶的表现。

阮郜·《阆苑女仙图卷》（局部）

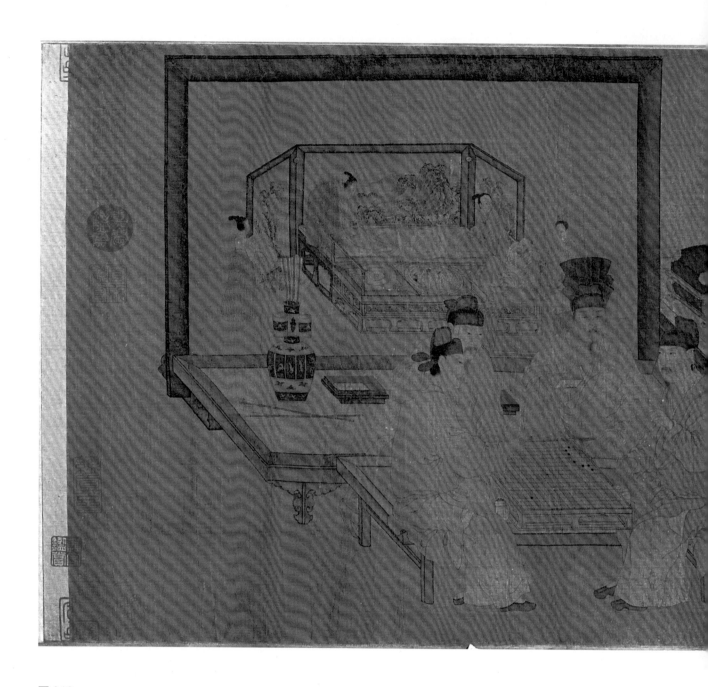

图 012

五代·周文矩·《重屏会棋图》·北京故宫博物院藏品

  绢本设色，高 40.3 厘米，宽 70.5 厘米。

  在会棋者之后的两屏间，有一老一少，正执壶点茶，茶盏带卷荷叶状的盖。

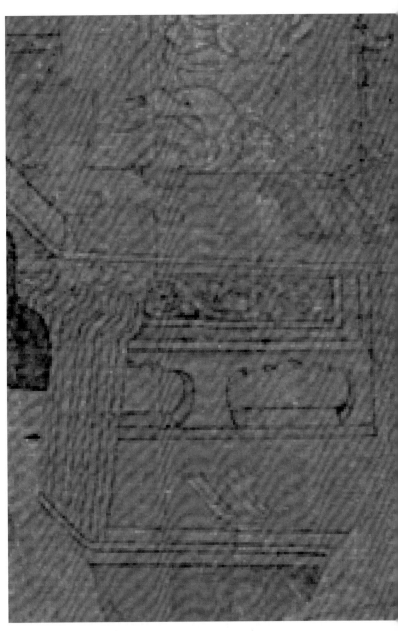

周文矩·《重屏会棋图》（局部）

图 013

五代 · 卫贤 · 《高士图卷》· 北京故宫博物院藏品

　　这是以奉茶为主题的"举案齐眉"！

卫贤·《高士图卷》（局部）

在河北、辽宁、北京及山西、陕西等地的辽金元壁画墓中，我国考古工作者发现了相当数量的《备茶图》。这些壁画填补了文献资料的空白，表明辽金元饮茶直接来源于唐代与宋代，以点注饼茶为主流，也兼顾了儒释道价值取向。例如宣化七号墓的《三教会棋图》和叶茂台七号辽墓《深山会棋图》。虽然这些墓主多为汉人官员，其《备茶图》《奉茶图》几乎与宋人没有二致，但从一个侧面表明唐宋茶道文化在这些地区这些民族之中有极大的影响力，在文化创造方面，汉人无疑走在契丹、室韦和女真党项族前面。

成书于 1269 年的审安老人《茶具图赞》，记载了标准的点茶器。

1. 用风炉或烤笼炙烤茶饼

2. 砧锥，用以碎茶饼

3. 茶槽子、碾子，用以碾茶

4. 石磨，用以碾磨茶

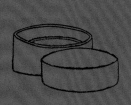

5. 竹罗，筛罗茶末

6. 棕刷，以清扫茶末

第二章

宋、辽、夏、金、元
时代茶文化
及茶画制作

Chapter Two
Tea Culture and Tea-Related
Painting Creation of the Song,
Liao, Xia, Kin and Yuan Dynasties

陶寶文

7.茶盏，用以点茶

滲提點

8.执壶，用以冲注，为
防烫手，有垫碗

竺副师

9.竹制茶筅，击拂茶汤，
制作白泡沫

胡員外

10.葫芦水瓢，用以舀水

漆雕秘阁

11.茶盏茶托，品茶

司职方

12.丝织茶巾，清理几
案、茶具

饮茶文化兴起于两晋，以杜育《荈赋》为标志；茶道文化始于李唐，以中唐时期约 761 年初稿完成并为天下所知、780 年前后正式定稿刊行的《茶经》为标志。中唐时期达到中国茶道文化的第一高峰：《茶经》的刊行使人们开始以更加科学的态度对待茶的种植、加工以及对茶具茶器的更科学更美观地设计与制造，人们以儒家文化为核心，以家国情怀为根本，以佛教的禅定与道家的羽化成仙为辅，赋予茶以多元的文化内涵。人们对茶、茶器、水性、品茗及环境品茗场所的审美逐渐成为一种综合的新的文化形式，从而自然而然地成为诗词歌赋的重要题材。

茶道，就是人们以烹茶饮茶为载体，将内心的审美与文化观念进行外在地自然地体现与表达，将内化于心的儒释道借由茶事而外化于行，进而将赋予茶的文化贯彻于生命存在的时时处处。茶成为综合文化的标志物，烹饮茶的茶艺与琴棋书画插花焚香技艺有机结合，形成宋代特色之一。

经过唐末五代的战乱，960

The tea-drinking-related culture marked by Du Yu's "Ode to Tea Planting and Drinking" sprang up in the Western and Eastern Jin Dynasties. The tea-ceremony culture began in Li Tang Dynasty, marked by *The Classic of Tea* which first draft was completed in 761 of the Middle Tang Dynasty, known to the public at that time and officially published around 780. The Middle Tang Dynasty reached the first peak of Chinese tea-ceremony culture. *The Classic of Tea* which made people started to have a more scientific attitude towards the tea planting, the tea processing and also the designing and manufacturing of the tea set. People began to give tea a diverse cultural connotation on the basis of Confucian culture as its core, feelings of home and native country as its main element, Buddhist meditation, Taoist ascending to heaven and becoming an immortal as its complementary element. A new cultural mode thus was cultivated via people's aesthetic appreciation of tea including tea leaf, tea set, quality of water, tea drinking, and tea-drinking environment, which naturally became an important subject of poetry.

The tea ceremony is a kind of natural presentation of people's aesthetic appreciation and cultural concept containing the connotation of Confucianism, Buddhism and Taoism with tea cooking and tea drinking behaviors. It turned out to be an indispensable part of everyday life all over the country. Tea became a symbol of comprehensive Chinese culture, drinking tea thus a demonstration of the art of Guqin playing, Chinese-chess playing, Chinese calligraphy and painting.

The Song Dynasty was formally established in 960 after the wars in the Late Tang Dynasty and the Five Dynasties thus unifying the core regions of ancient China. Chinese culture entered into its mature period by summarizing its success and failure of the Tang Dynasty, hence Chinese tea ceremony

年宋王朝正式建立，并进而实现中国古代核心区域的统一。在总结汲取唐代成败的基础上，进入中国文化的成熟时期，茶道文化也展示出时代性特点。

（一）以苏廙《十六汤品》为代表的晚唐点茶成为宋代茶道的主流。以点茶为主要形式的茶汤斗试一时从福建传到关中，进而传到宋东西二京。点茶是赵佶（1082—1135）《大观茶论》、蔡襄《茶录》等的主要论述点。由于气温下降，贡茶区由长江流域向闽江流域转移。建州北苑茶区成为主要名茶产地，制茶更为精细。北宋宫廷用茶可能成为中国历史上成本最高的茶。

（二）以点茶器为核心的茶执壶、斗笠形茶盏成为茶器主流。以南方的建盏备受追捧，而北方耀州窑为代表的青瓷器则与后来的景德镇青花瓷媲美，成为考古出土分布最广的瓷器，也是世界各地博物馆收藏茶盏最多的瓷窑之一。当然由于地域原因，辽代宣化墓出土茶器则以定窑白瓷为主。在宋徽宗君臣的提倡下，建盏难得，仿制品则到处都有，江西吉州窑、关西耀州窑

culture also embodied the characteristics of the time.

Firstly, Su Yi's *Sixteen Tea Liquors* showed the method of cooking tea by pouring boiling water into cup with grinded tea dust popular in the Late Tang Dynasty became the mainstream of the Song Dynasty tea ceremony. It seemed that tea liquor competition was introduced from Fujian to Guanzhong of Shaanxi, then to the eastern and western capitals of the Song Dynasty. The method of cooking tea by pouring boiling water into cup with grinded tea dust was the the main point of the book *Discussion on Tea* by Zhao Ji, emperor Huizong of the Song Dynasty, and the other book *Tea Records* by Cai Xiang. The royal tribute tea planting area transferred from the Yangtze River valley to the Minjiang River valley as the temperature dropped. Beiyuan tea planting area in Jianzhou Prefecture became the main tea producing area. Tea was more delicately made at that time and the royal tribute tea of the Northern Song Dynasty could be the highest-cost tea in Chinese history.

Secondly, handled ewer and calyx played the key role in the tea-cooking by means of pouring boiling water into cup with grinded tea powder. The calyx made in Jianzhou kiln in south China and celadon tea set made in Yaozhou kilnin north China were popular and comparable to blue and white porcelain tea set made in Jingdezhen kiln later, were also the most widely discovered porcelain tea set in archaeological excavation and collected by museums around the world. Those tea set found in Liao-Dynasty tombs at Xuanhua were mainly of white porcelain from Ding-kiln. Porcelain calyx made in Jianzhou was rarely seen in the Huizong Reign of the Southern Song Dynasty but its replicas especially those produced in Jizhou kiln in Jiangxi Province and Yaozhou kiln in Guanxi were easily got every where. The purple clay

第二章 宋、辽、夏、金、元时代茶文化及茶画制作
Chapter Two Tea Culture and Tea-Related Painting Creation of the Song, Liao, Xia, Kin and Yuan Dynasties

057

都大量仿制建盏。

在制造陶瓷过程中，紫砂壶也出现在宋朝。

（三）对茶艺的追逐已经普及到各个阶层。范仲淹《和章岷从事斗茶歌》将新茶上市、各家隆重斗试质量高下的情形描述得淋漓尽致："胜若登仙不可攀，输同降将无穷耻。"这与唐代以皇宫贵族、士大夫、寺宇道观为主体的茶事不同，茶区全民参与，达到狂热程度。敦煌文书《茶酒论》成为民间对茶道文化的理解与称赞。

（四）这一时期茶人对茶道文化、茶人精神有了更为深刻的表达。《大观茶论》对茶道文化净化社会风尚、营造清廓和雅致风民俗寄予很高希望。而唐宋八大家之一的苏轼则以一部《叶嘉传》，树立了一个资质刚健、不畏艰险、直面敢言的君子形象。无疑这是中国士人心目中的茶人标准。苏轼可谓宋代，也是古代中国茶人的杰出代表：传统文化修养极为丰富，诗书画样样精通，为人爱憎分明，性格却温婉和雅，气质清爽倜傥，屡屡遭贬，但即使远到海南，依然笑对人生。

teapot appeared in the Song Dynasty.

Thirdly, people from all walks of life tended to learn and practise the tea art in the Song Dynasty. The poem "Tea Competition" by Fan Zhongyan described vividly the situation of new tea appeared on the market and tea-quality competition such as the following two sentences: It is as hard to win the competition as becoming an immortal, one will be ashamed just like a surrendered enemy general to lose it. This is different from the tea activities participated by the imperial palace aristocrats, scholars, officials, Buddhist monks and Taoist monks in the Tang Dynasty. One of the Dunhuang Documents "Views on Tea and Liquor" represented a popular folk understanding and praise of the tea ceremony culture.

Fourthly, a deeper expression on the tea ceremony culture and the tea man spirit took its shape in this period. Discussion on Tea gave a high hope for the purification of social customs, the creation of harmonious political atmosphere, and folk custom via the tea ceremony culture. Su Shi set up an image of a brave-to-speak-his-mind gentleman in his book *Biography of Ye Jia* making Ye Jia undoubtedly a standard tea man. Su Shi himself was also an outstanding representative of Chinese tea man of the Song Dynasty with extremely rich knowledge of traditional Chinese culture, multiply skills of poem, calligraphy and painting, gentle and elegant character, personality of a clear-cut stand on what to love and what to hate, fresh and charming temperament. He experienced repeatedly degradation in officialdom, being banished even as far as to Hainan Island, but he still kept smiling to life.

The range of aesthetic appreciation towards the tea art had been broadened by adding the procedure of tea grinding into

人们对茶艺的审美范围更加扩大：磨茶进入茶道过程中。斗试茶汤色泽、滋味成为茶事成败的关键点。"茶百戏"则成为点茶技艺的娱乐内容，最早可从唐代诗歌中发现其端倪。

（五）宋代茶马贸易继续发展，茶的输出量更加扩大。澶渊之盟规定每年供给辽 25 万斤茶叶，这是不小的数量，同时还向西夏、金、元输出茶叶。宋代茶马贸易制定严格政策，但是得不到切实执行。相互以茶礼待使者，成为宋、辽、夏之间约定成俗的宫廷礼仪，是最为通行的外交礼法。而宋是宫廷茶道礼仪的中心，因为宋朝是唐、五代茶文化的直接继承者，为一种高雅文化在辽、夏、金的普及发展提供资源。从辽代宣化墓等考古资料可以看到：辽代茶事比宋朝更显热闹喧哗，赏花插花、茶器具尚白色，这可能与契丹的传统和地处河北定窑瓷文化圈有关。辽、夏、金茶文化更多一些佛教色彩。而在民间，辽、夏、金的酒文化气氛要远远比茶文化浓厚。在墓室壁画中，存酒瓶几乎成为展示墓主生活必不可少的道具。

the tea ceremony. The color and flavour of tea liquor were the key points in the tea competition. "Tea drama", on the other hand, had become the entertainment content of skill of tea cooking by pouring boiling water into cup with grinded tea dust, whose earliest records can be found in Tang-Dynasty poetry.

Fifthly, the tea-horse tradein the Song Dynasty continued to develop, and the output of tea expanded. The Song court had to supply 125000kg of tea to the Liao court every year under the cause in the Oath of Chanyuan, but also exported tea to Xixia (Western Xia Regime) court, Kin court, and Yuan court. A strict tea-horse trade policy had been formulated in the Song Dynasty but had not been implemented effectively. Receiving the royal emissaries with tea ceremony had become the diplomatic etiquette during the Song, Liao, and Western Xia dynasties. The Song court, as a direct successor of the tea culture of the Tang Dynasty and the Five Dynasties, played a leading role in the royal tea ceremony etiquette, also for the popularization and development of a elegant culture within the territory of the Liao, Western Xia, and Kin regimes. We can see from archaeological data such as those found from Liao-Dynasty tombs at Xuanhua that tea activities in the Liao regime were more lively and noisy than that in the Song regime. Liao people loved flower arrangement and white-porcelain tea set which might owes to its own tradition and geographical location—in the cultural circle of Ding-kiln of Hebei Province. The tea culture of the Liao, Western Xia, and Kin regimes contained more Buddhist elements while liquor culture dominated the folk life. Liquor bottle had almost been a daily-life necessary in those years seen from tomb burial objects.

Sixthly, the tea culture research prospered in the Song

第二章 宋、辽、夏、金、元时代茶文化及茶画制作
Chapter Two Tea Culture and Tea-Related Painting Creation of the Song, Liao, Xia, Kin and Yuan Dynasties

059

（六）宋代是茶学繁荣的时代。有陶穀（903—971）《茗荈录》、丁谓（966—1037）《北苑茶录》、周绛《补茶经》、叶清臣（1000—1049）《述煮茶泉品》、刘异《北苑拾遗》、蔡襄《茶录》、宋子安《东溪试茶录》、黄儒《品茶要录》、沈括（1031—1095）《本朝茶法》、赵佶（1082—1135）《大观茶论》、唐庚（1071—1121）《斗茶记》、熊蕃《宣和北苑贡茶录》、桑庄《续茶谱》、赵汝砺《北苑别录》、审安老人《茶具图赞》等。

我们也看到有元代宰相耶律楚材等人的咏茶诗，有赵孟頫等的茶画，但相对而言，对茶艺的追求比唐宋略为逊色，而对茶道思想、茶道理念的探索几乎停滞。民间受到游牧民族的影响，饮茶有所收缩。茶马贸易政策也随着蒙古入主中原也自然消亡。

Dynasty with these evidences: *Tea Tasting* by Tao Gu (903–971), *Record on Tea from Beiyuan* by Ding Wei (966–1037), *The Supplementary Classic of Tea* by Zhou Jiang, *Spring Water for Cooking Tea* by Ye Qingchen (1000–1049), *Complementary Records on Tea from Beiyuan* by Liu Yi, *Tea Records* by Cai Xiang, *Identifying the Origin of Tea through Tasting* by Song Zi'an, *Record of Tea Making* by Huang Ru, *Tea-relevant Laws and Regulations* by Shen Kuo (1031–1095), *Discussion on Tea* by Zhao Ji, *Tea Competition* by Tang Geng (1071–1121), *Xuanhe-Reign Tribute Tea from Beiyuan* by Xiong Fan, *Extended Tea Manual* by Sang Zhuang, *Outline Record on Tea from Beiyuan* by Zhao Ruli, and *Catalogue of Tea Set* by Shen'an Laoren.

Although we can still see the tea-related poems by Yelv Chucai, the prime minister of the Yuan Dynasty, tea-relevant paintings by Zhao Mengfu, etc. Today, relatively speaking, the pursuit of the tea art in this period was slightly inferior to that of the former dynasties, the exploration of tea-ceremony thoughts and concept almost stagnated. The social needs for tea drink contracted under the influence of the nomadic people and the tea-horse trade policy also vanished naturally as the Mongolians conquered the central plains of China.

图014

宋·王诜（1048—1104）·《绣栊晓镜图》

　　画中女主人公晨起梳妆打扮后照镜，白色卷荷叶盖茶盏放于镜架之下，
热茶一杯是一天生活的开始。

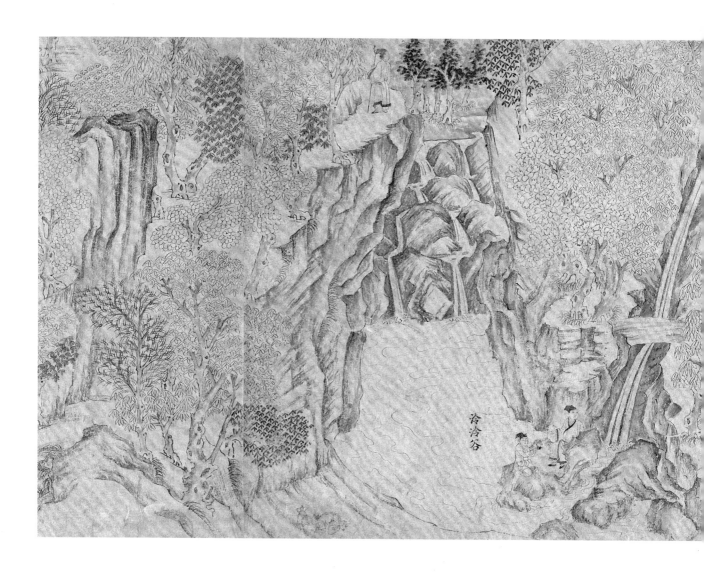

图 015

宋·李公麟（1049—1106）·《李龙眠山庄图·延华洞》·台北故宫博物院藏品

　　宋代是我国封建社会的成熟期。陈寅恪等史学家认为，汉唐以来的中国文化在宋代才真正达到成熟，并日臻完善。
在茶道文化诸方面也是如此。在茶学研究方面，以蔡襄（1012—1067）的《茶录》和宋徽宗赵佶（1082—1135）的《大
观茶论》为代表。蔡襄是中国茶文化史上继陆羽、卢仝以后又一位杰出茶人。他的《茶录》上篇论茶，下篇论器偏重
于茶艺方面，在茶道思想方面没有着墨。而徽宗的著作在序言中开宗明义："本朝之兴，岁修建溪之贡，龙团凤饼，
名冠天下，而壑源之品，亦自此盛。延及于今，百废俱举，海内晏然，垂拱密勿，幸致无为。……故近岁以来，采择
之精，制作之工，品第之胜，烹点之妙，莫不咸造其极……天下之士，励志清白，竞为闲暇修索之玩，莫不碎玉锵金，
啜英咀华，较箧笥之精，争鉴裁之妙，虽下士于此时，不以蓄茶为羞，可谓盛世之清尚也。"这既是对当时茶道文化
的肯定，同时也是一种文化引导。除了人们熟知的宋辽、宋金和约中，宋朝每年要向这些国家提供数量很大的茶叶，
在接待外国使臣、皇帝会见大臣、视学等活动中都要赐茶。而茶在徽宗朝达到历史新高度。

李公麟·《李龙眠山庄图·延华洞》（局部）

　　主人（李龙眠）坐于巉石之上，谛听飞流激水。书童倾身向他奉茶，而他似乎融入山水之中，对书童的到来很不经意。

但奉茶是书童每天的必修课！

第二章 宋、辽、夏、金、元时代茶文化及茶画制作
Chapter Two Tea Culture and Tea-Related Painting Creation of the Song, Liao, Xia, Kin and Yuan Dynasties

063

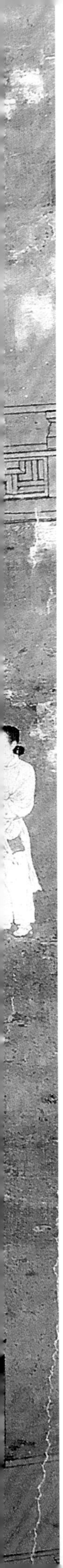

图 016

北宋·赵佶（1082—1135）·《文会图》

　　《文会图》具体描绘了北宋时期文人雅士品茗雅集的一个场景。地点应该是一所庭园或者皇帝御花园，旁临曲池，石脚显露。四周栏楯围护，垂柳修竹，树影婆娑。树下设一大案，案上摆设有果盘、酒樽、杯盏等。大案前设小桌、茶床，小桌上放置酒樽、菜肴等物，一童子正在桌边忙碌，装点食盘。茶床上陈列茶盏、盏托、茶瓯等物。左烧水炉，茶桌、茶床共同构成事茶主要器物茶案。画面下方五人，从左到右分别是：第一人坐于木盖上喝茶；第二人站在炉前等候（准备与第三人即最中间者进行点茶）；第三人右手正从茶末罐里勺取茶末、左手端茶托茶盏；第四人正在端盘侍立，正在等第二、三人点好茶后去奉送，其幞头、袍服极讲究，有品级；第五人正在归置茶床。床旁设有茶炉、茶箱等物，炉上放置茶瓶，炉火正炽，显然正在煎水。有意思的是第一人，即那位坐着的青衣短发小茶童，也许是渴极了，他左手端茶碗，右手扶膝，正在品饮。他的职责也许是兼顾品鉴茶汤滋味。与陆羽《茶经》中介绍的茶具茶器不同，在这场茶事中多了一个带鼎足的盛水铁锅，应是贮清水的器物。还有一个组合：放在炉边的带盖短直腹罐、敞口盆、小方几。罐中很可能是盐，但作为茶末罐可能性更大。白面黑框的盝顶四角匣，可能是仅仅装放茶盏、茶托的都篮（《大观茶论》没有具列）。图中右上有赵佶亲笔题诗："题文会图：儒林华国古今同，吟咏飞毫醒醉中。多士作新知入彀，画图犹喜见文雄。"图左中为"天下一人"签押。左上方另有蔡京题诗："臣京谨依韵和进：明时不与有唐同，八表人归大道中。可笑当年十八士，经纶谁是出群雄。"

　　画中前方是一组备茶备酒图。左边着黄衣者在炉子上用执壶烧水；右边小床一黑一绿衣者大概备酒。大案子下边有瓶酒。欣赏这幅画还应该注意的细节是，前边五人备茶酒中，最中间的一个正在左手端盏右手用勺从罐中舀茶，虽然人较多，但还是一人一杯地点茶。赵佶《大观茶论》是宋代茶道的巅峰之作，提出了中国茶道文化的最高精神旨趣："祛襟涤滞，致清导和；冲淡简洁，韵高致静"，"天下之士，励志清白，啜英咀华，可谓盛世之清尚也"。

第二章　宋、辽、夏、金、元时代茶文化及茶画制作
Chapter Two Tea Culture and Tea-Related Painting Creation of the Song, Liao, Xia, Kin and Yuan Dynasties

065

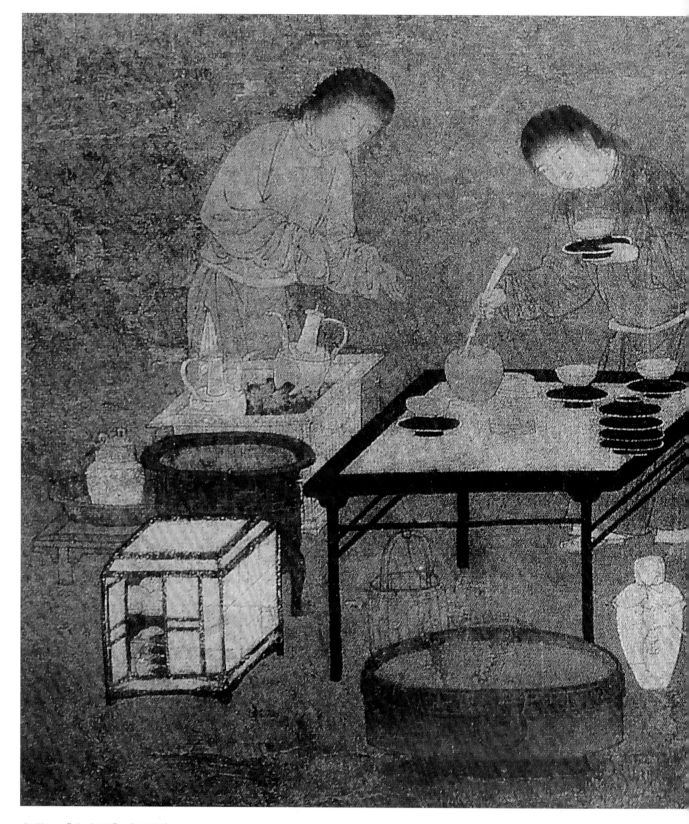

赵佶·《文会图》（局部）

　　茶桌前边是大口深腹木盆，应为审安老人《茶具图赞》中的"水方"。右边一组为"备酒图"：温酒器、勺茶杓、酒瓶密封。带覆斗笠盖的斜腹筒形器，应是滤酒器。画面最左侧是四腿方几，上置一铁盆，盆内有一带盖罐，应该是贮茶罐，没有见到碎茶砧板和碾槽，可能是末茶成品。替代陆羽"具列"的是方形漆边方案。另一个值得关注的信息是：茶盏茶托与酒床上的酒杯酒台子在造型及颜色搭配上有明显差别。酒壶与茶执壶差别不是十分明显。茶壶

盖简约、器形挺秀，酒壶盖复杂器形浑圆敦实。茶壶流
嘴稍微平折。这里没有见到茶执壶的衬手碗。酒壶带有
温碗。

第二章 宋、辽、夏、金、元时代茶文化及茶画制作
Chapter Two Tea Culture and Tea-Related Painting Creation of the Song, Liao, Xia, Kin and Yuan Dynasties

067

图 017

宋·张择端（1085—1145）·《清明上河图》·北京故宫博物院藏品

　　清明上河，实指清明节前后，人们到汴河两岸的汴京城游览、赏春、逛集市，
茶肆、酒楼分布比较密集，说明茶已深入普通百姓生活。

图 018

宋·苏汉臣（1094—1172）·《罗汉图》册页·日本人私藏

原画高 25.6 厘米，宽 24.4 厘米。

高僧跏趺坐于竹椅之上，一侍童点火煮水，一侍童骑坐于几床之上磨茶，另有一侍

童用盘盛茶托、茶盏、带勺茶罐，准备来点茶。

图 019

宋·苏汉臣·《靓妆仕女图》·美国波士顿艺术博物馆藏品

团扇，绢本设色，高 25.1 厘米，宽 26.7 厘米。

夫人梳妆台上有大铜镜、梳妆盒（套装）、茶杯、插花瓶。清晨
饮茶大约是贵族妇女的普遍习惯。

第二章 宋、辽、夏、金、元时代茶文化及茶画制作
Chapter Two Tea Culture and Tea-Related Painting Creation of the Song, Liao, Xia, Kin and Yuan Dynasties

071

图 020

宋·苏汉臣·《长春百子图·荷庭试书》（局部）

从苏汉臣《长春百子图》可以初步判断：北宋已经出现紫砂壶。紫砂器可能源于瓷器的素烧而已，由于徽宗君臣大力提倡点茶斗试，以建州兔毫盏为最佳茶器，形成对新兴的紫砂壶某种程度的掩盖或客观上的压制，才使得人们将紫砂的最初使用时间后置于明代。

图 021

宋·苏汉臣·《文姬归汉图卷·第十八拍：归来故乡》（局部）

图左下方是两人坐在茶肆品茶。

图 022

宋·李公麟（1049—1106）·《会昌九老图卷》·北京故宫博物院藏品

　　此图描绘了唐会昌五年（845）三月二十四日九位退休老人白居易、胡杲、吉皎、郑据、刘真、卢真、张浑、李元爽、释如满等九位老人相聚洛阳履道坊白居易居所欢聚"尚齿"之会，茶酒相间，诗画共赏的情景。在他们会心相聚的同时，隔壁有人在专门的茶间为他们备茶。茶盏摞得很高，茶床较大，两个执壶同时放在炉上。这一切说明主人的确是一位"别茶人"，且茶友颇多。谈笑皆鸿儒，往来茶友多。在京为官时，蒙山新茶寄来，诗人"渭水煎来始觉珍"。为避免卷入牛李党争的旋涡，乐天赴洛阳任东都太子宾客，以"香山居士"名号，结交僧俗道人，诗友雅士往来者多。在他的故居也发现一些茶器文物，这绝非偶然，而且与他一生喜爱茶事息息相关。他不仅长期在京为官，宪宗朝拜翰林学士，历左拾遗、中书舍人，知制诰，还在茶区任江州（今江西九江市）司马、杭州刺史（正三品）；文宗朝任刑部侍郎，后转东都留守。武宗会昌初致仕（退休）。《新唐书·白居易传》记载他："卜居履道里，与香山僧如满等结净社。疏沼种树，构石楼，凿八节滩，为游赏之乐，茶铛酒杓不相离。"白居易与李白、杜甫相比，一生经济比较优渥，不为禄米所忧扰。其大量的诗篇成为唐代历史的动态画面。他也是晚唐北方诗人写茶诗最多的人。史书记载他骨鲠敢言，究习佛理较深，在皇帝降诞节的三教辩论中代表儒教与释、道二教较量长短高下。是与茶圣陆羽、亚圣卢仝、诗僧皎然齐名的大茶人，而他的经历更丰富、社会地位更高、更接近最高政治核心。

　　李公麟《会昌九老图卷》形象地再现了唐代中国茶道的基本特征：书画助兴、诗词激扬茶道思想，描摹茶道之美，与酒相携；重在品饮，重在饮茶后的精神升华与文思诗艺的会心碰撞。

第二章 宋、辽、夏、金、元时代茶文化及茶画制作
Chapter Two Tea Culture and Tea-Related Painting Creation of the Song, Liao, Xia, Kin and Yuan Dynasties

073

图 023

宋 · 佚名 · 《斟酒图》· 台北故宫博物院藏品

　　人物册页，图中应为斟酒图，首先，喝茶的白瓷盏托还在方几上的网罩之内，还未启用；其次，从侍童用左手掌心托壶底的情形看，应该是在倒酒。因为点茶执壶直接在火上煮水，一般用瓷碗垫底，是不能用手直接接触执壶底部的，以防烫伤。

　　原画名为《斟酒图》。但我们却发现：酒瓮已经上封，而煮茶的炉子和长柄已经取下青纱罩子，茶事即将开始。

图 024

宋·刘松年（1155—1218）·《博古图》·台

北故宫博物院藏品

　　原高 128.3 厘米、宽 56.6 厘米。

　　像其他古画一样，这幅画中的煎茶侍童也

被安排在一个角落。但侍童煮水扇风的动作画

得极为细腻，他显得极为专注、沉静。

第二章　宋、辽、夏、金、元时代茶文化及茶画制作
Chapter Two Tea Culture and Tea-Related Painting Creation of the Song, Liao, Xia, Kin and Yuan Dynasties

075

图 025

宋·刘松年·《斗茶图》·台北故宫博物院藏品

绢本设色，高 57 厘米，宽 60.3 厘米。

这表现的是斗茶已经结束，双方正在谈论，互相切磋。值得注意的是，炉子上的壶身类似紫砂壶造型。但好像带把手，唐长沙窑多有其器。

斗茶，也叫茗战。冯贽《记事珠》："茗战，建人斗茶为'茗战'。"斗茶就是以点茶形式，比试不同的点茶者的技艺、茶质。《大观茶论》是对点茶道设立法则及美学标准的经典文献，也是徽宗皇帝对《茶经》烹茶道以来，为点茶道树立新法，我们全文录取：

"点茶不一。而调膏继刻，以汤注之，手重筅轻，无粟文蟹眼者，调之静面点。盖击拂无力，茶不发立，水乳未浃，又复增汤，色泽不尽，英华沦散，茶无立作矣。有随汤击拂，手筅俱重，立文泛泛，谓之一发点。盖用汤已故，指腕不圆，粥面未凝，茶力已尽，云雾虽泛，水脚易生。妙于此者，量茶受汤，调如融胶。环注盏畔，勿使浸茶。势不欲猛，先须搅动茶膏，渐加击拂，手轻筅重，指绕腕旋，上下透彻，如酵蘖之起面。疏星皎月，粲然而生，则茶面根本立矣。第二汤自茶面注之，周回一线，急注急止，茶面不动，击指既力，色泽渐开，珠玑磊落。三汤多寡如前，击拂渐贵轻匀，周环旋复，表里洞彻，粟文蟹眼，泛结杂起，茶之色十已得其六七。四汤尚啬，筅欲转稍宽而勿速，其真精华彩，既已焕然，轻云渐生。五汤乃可稍纵，筅欲轻匀而透达。如发立未尽，则击以作之；发立已过，则拂以敛之。结浚霭，结凝雪，茶色尽矣。六汤以观立作，乳点勃然，则以筅著居，缓绕拂动而已。七汤以分轻清重浊，相稀稠得中，可欲则止。乳雾汹涌，溢盏而起，周回凝而不动，谓之咬盏。宜匀其轻清浮合者饮之，《桐君录》曰，"茗有饽，饮之宜人，虽多不力过也。"

我们理解的"发立"是指茶粉与沸水迅速融合；击，是用竹筅拍打，拂，旋转，两个动作相互配合；粟纹、蟹眼是一个意思，就是小泡沫；茶力，指茶粉遇水挥发的力度和持久力，比如小麦面比玉米面挥发力好，而石灰更好；粥面，指粥一样的一层凝结物；水脚，水痕，泡沫破灭，水痕在茶盏边出现；饽，意思与粥面一样，有点像我们今天啤酒的一层泡沫。总之是追求泡沫多而持久，咬盏有力，先出现泡沫，最后出现水痕，为点茶动作比试的胜利者。

其次是茶的味道，现在有的人看到《大观茶论》特别注重

上述的动作和对泡沫及其持久力，认为那是斗茶的主要内容。其实一切都是为一盏味道上佳的茶，为了品饮。徽宗明确指出：

味，夫茶以味为上。香甘重滑，为味之全。惟北苑壑源之品兼之。

再次是香。

香，茶有真香，非龙麝可拟。要须蒸及熟而压之，及干而研，研细而造，则和美具足。入盏则馨香四达。秋爽洒然。或蒸气如桃仁夹杂，则其气酸烈而恶。

最后是追求茶汤之色。赵佶认为：

色，点茶之色，以纯白为上真，青白为次，灰白次之，黄白又次之。天时得于上，人力尽于下，茶必纯白。天时暴暄，芽萌狂长，采造留积，虽白而黄矣。青白者蒸压微生。灰白者蒸压过熟。压膏不尽，则色青暗。焙火太烈，则色昏赤。"

这是点茶道基本法则。点茶比试，一看泡沫高低出现早晚、持久力；二尝滋味；三嗅其香，反对添加人工之香；四看茶汤颜色。四者缺一不可，比的是综合分。

刘松年，南宋孝宗、光宗、宁宗三朝的宫廷画家。宦居于著名的钱塘清波门，故有"刘清波"之号。是南宋茶事画的代表。这幅《斗茶图》告诉我们许多点茶道历史信息。

一、时序和斗茶场地。清明前后的室外，很可能是范仲淹诗歌中的茶园不远处。看起来左边的点茶手右手执执壶，实际上是比试结束后在交流、比画。点茶是两个人共同完成：一人点水，一人持碗击拂。

二、画面情形。画面右边的已经尝过了自己和对方的茶汤，等待画面左边的最后品尝结果。可以看到每家的具列上各有两个茶盏，一只尝自家茶，另一只尝对方茶。属于没有裁判的友谊斗试。犹如较量武功，高下清楚，是从茶道审美的要素出发的比试。

三、使用的道具——制茶之具与饮茶之器。有具列（桌形竹架）、都篮（装放茶器）、茶风炉、木炭筐及蒲扇（双方公用）、敞口直腹白瓷圈足茶碗（未见托和斗笠盏）。出现了带流横鋬壶，很可能就是紫砂壶。同样式的在唐代安史之乱后的长沙窑多有出现，也在南海沉船中出现。并出现了冰裂纹开片较大的汝窑瓷罐。有圆盖束颈四平顶溜肩收腰瓷罐，可能是贮存茶末用，斗茶好像从点茶开始，唐代的一些程序被宋人简化了。

第二章 宋、辽、夏、金、元时代茶文化及茶画制作
Chapter Two Tea Culture and Tea-Related Painting Creation of the Song, Liao, Xia, Kin and Yuan Dynasties

077

图 026

宋·刘松年·《市井斗茶图》

　　炭炉不可缺少，都篮、具列都简化了，只有执壶和茶杯。看来，比的主要是味道！

图 027

宋·刘松年·《磨茶图》

　　在盆里点茶，供较多的人同时品饮。

图 028

宋·刘松年·《十八学士图》（局部）

　　较为全面地展示了两宋文人的高雅生活。茶是他们不可或缺的爱好！他本人也是中国茶道文化应该关注的历史人物。在他的不少画作中都有茶道元素。

　　画中描绘夏天，四个学士在一棵古柳之下，专心致志地下围棋，五个侍童为之服务：一个挥扇驱蚊蝇，一个扛持拂尘，一个侍立，两个正在点茶。

图 029

宋·刘松年·《六人斗茶图》

　　左边一组的斗茶士有力的左手回环紧握小小的茶盏，右手将金属大口细颈执壶提起，全身力气运发于平直的流口。他身后的副手正在添加木炭，右边一组仔细品尝观察。三人分工：一人持杯执壶点茶、一人烧水、一人将舀好茶末的杯子递给点茶者。

图 030

仿刘松年·《撵茶图》·台北故宫博物院藏品

　　原画高 16.5 厘米，宽 22.8 厘米。

图 031

宋·刘松年·《补衲图》·台北
故宫博物院藏品

　　原图高141.9厘米,宽59.8厘米。
　　一个高僧穿针补衲服,正用牙咬
断线头,一个仔细端详,两个侍者正
在准备点茶。

第二章 宋、辽、夏、金、元时代茶文化及茶画制作
Chapter Two Tea Culture and Tea-Related Painting Creation of the Song, Liao, Xia, Kin and Yuan Dynasties

081

图 032

宋·刘松年·《四景山水画卷》·北京故宫博物院藏品

桃李芬芳之春天傍晚，主人回归，侍者烧水备茶。

图 033

宋·李嵩（1166—1243）·《罗汉图》·台

北故宫博物院藏品

原图高 104 厘米、宽 49.5 厘米。

罗汉坐于榻床之上，一沙弥左手持觥

形杯，右手用匙添加东西，或香粉，或茶汤，

另一小侍者蹲地抚弄时花。

第二章 宋、辽、夏、金、元时代茶文化及茶画制作
Chapter Two Tea Culture and Tea-Related Painting Creation of the Song, Liao, Xia, Kin and Yuan Dynasties

图 034

宋·佚名·《春游晚归图》

　　此图原载《纨扇画册》（见《石渠宝笈三编》）。无作者姓氏。画面甚见渺远，充溢了春天的气息。
一老者策骑缓行，几个侍从各携椅、凳、食盒之属，可作一幅宋朝风俗画观。虽然看不到烹茶场面，也看
不到茶具、茶器，但可以为人们研究野外茗事提供参考。

图 035

宋·马远（1190—1279）·《西园雅集图》·美国纳尔逊·艾京斯艺术博物馆藏品

绢本淡设色。

宋代有很多绘画世家，其中最为出名的恐怕要属山西的马家。自北宋后期的马贲开始，马家先后有五代人在皇家画院供职。马氏家族形成了一个庞大的画室或作坊，在这个作坊内雇用了助手、管理者或代理人，或许还有绘画材料的生产者及裱画匠。元祐元年（1086），苏轼兄弟、黄庭坚、李公麟、米芾、蔡肇等十六位名士，于驸马王诜宅邸西园集会。马远据此所绘，长卷共分四段。此为其中一段，绘米芾挥毫作书，诸文友或立或坐，凝神围观。

在主体位置之外，有茶几、茶器。在文人雅集活动中，茶汤是必备的饮品。

图 036

宋·佚名·《夜宴图》·美国私人收藏品

绢本设色，高 26.2 厘米，宽 160 厘米。

酒醉以茶来解。明显看出酒盏小于茶盏。这是唐宋以来"茶酒间进"习俗的延续。人物衣冠、画风有南宋特色。

《夜宴图》（局部）

这与图 035 应该是一幅画。因来自不同地方，便分为两图。本图可视为图 035 的局部放大。茶盏有青釉、白釉两种。茶托子有红、黑两个颜色（白丧事忌讳用红托子）。而酒盅没有台子，只有一个圈足盘。红烛高照，有的人已经跟跄离席，有的人酒醉，穿靴准备离席，有的人呼呼大睡。

第二章 宋、辽、夏、金、元时代茶文化及茶画制作
Chapter Two Tea Culture and Tea-Related Painting Creation of the Song, Liao, Xia, Kin and Yuan Dynasties

089

图 037

南宋·陈清波（1253—1258）·《瑶台步月图》·美国弗利尔博物馆藏品

　　作者为南宋理宗画院待诏。高 35.8 厘米、宽 35.9 厘米，从手捧茶盘情形看，被
定为南宋作品。

图 038

宋·张路·《风雨轴》·北京故宫博物院藏品

　　风吹树摇，雨下流急，一片急促景象；亭中主人独品香茗，一派宁静，愈显茶的清心宁虑
的妙理。

图 039

宋·佚名·《斗茶图》

　　清人摹本，绢本设色。

　　这是场面不大、倚山临水、紧张而又热烈的三方比试。三套都篮具列，三套风炉横柄执壶。

每人都有五六个茶（叶）末罐。三人都很自信、从容。侍者也很幽静。

第二章 宋、辽、夏、金、元时代茶文化及茶画制作
Chapter Two Tea Culture and Tea-Related Painting Creation of the Song, Liao, Xia, Kin and Yuan Dynasties

093

图 040

南宋·夏圭（1180—1230）·《雪堂客话》单页·北京故宫博物院藏品

  作者与马远同时，宁宗时任画院待诏。高 28.2 厘米，宽 44.5 厘米。

  山松、古树、竹林连同庭院房舍，笼罩在雪的世界之中，夜色朦胧。客主相对，香茗散发出阵阵清香。远方的客人不惧风大路遥，踏雪来访，是知音相思，是学有心得，或是穷究天理而遇山重水复不见出口，抑或人生多舛，且进一盏热茶，再听他娓娓道来。万籁俱寂，整个世界属于芬芳的茶香，属于情感的交流、思想的碰撞。中国茶道，与生活同在，与友情同在，与真理同在。茶是朴素的，它是生活的本来样貌；茶是深刻的，或究天人之理，或求治国之道；茶也是神圣的，它照映茶人的生命。

夏圭·《雪堂客话》单页（局部）

第二章 宋、辽、夏、金、元时代茶文化及茶画制作
Chapter Two Tea Culture and Tea-Related Painting Creation of the Song, Liao, Xia, Kin and Yuan Dynasties

095

图 041

南宋·陈居中·《文姬归汉全图》·台北故宫博物院藏品

　　宁宗嘉泰年间（1201—1204），作者为画院待诏。图中明显用执壶注水于直筒杯中，是点茶动作和器具。

图 042

南宋·仿周文矩·《饮茶图》

　　可能是南宋人假借五代周文矩之名而绘制的。画中一人端茶器，一人捧茶盘。中间
主人站立，后一侍女手捧锦囊食盒，可能是茶点。

第二章 宋、辽、夏、金、元时代茶文化及茶画制作
Chapter Two Tea Culture and Tea−Related Painting Creation of the Song, Liao, Xia, Kin and Yuan Dynasties

097

图 043

南宋·《唐人宫苑图卷》·北京故宫博物院藏品

    这里是唐九成宫、华清宫,还是翠微宫,我们不得而知,但其雅致的环境,
确是唐人品茗相聚的理想环境。

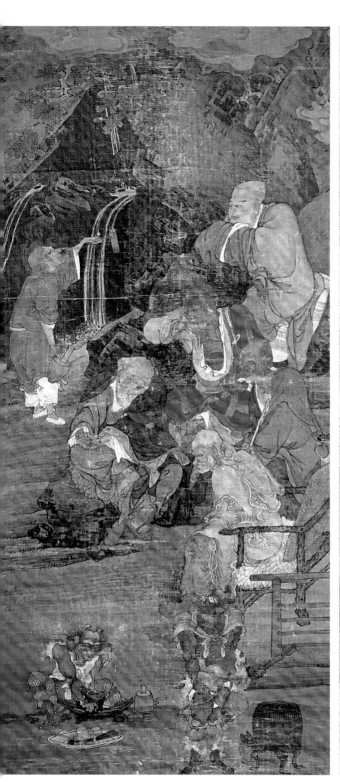

图 044

南宋 · 义绍 · 《五百罗汉图 · 备茶》

  唐天祐元年（904）中元节，有十六罗汉显化于宁波东钱湖青山顶，罗汉信仰遂兴。至南宋义绍住持东钱湖惠安院时，他邀请周季常、林庭珪两位画师用时 10 年绘制《五百罗汉图》，并供养于惠安院内。后来义绍以"大千世界佛日同辉"为旨，将百幅《五百罗汉图》赠予日本求法僧。这些画作先在镰仓寿福寺、藏箱根早云寺供养，1590 年移至京都丰国

寺，再至奈良大德寺。途中有6幅遗失，日本僧人木村德应在1638年补齐。1895年10幅作品转给美国波士顿美术馆、2幅作品转给华盛顿弗利尔美术馆收藏。所见图我们命名为《五百罗汉图·备茶》，近年曾参加国家文物局主办、浙江省博物馆承办的展览。图中下方为力士用金法槽碾茶、罗茶；侍童用执壶点注，罗汉端杯。与刘松年《磨茶图》《补衲图》联系思考，宋代碎茶后有三种方法将饼茶加工为茶粉（茶末）：石磨磨茶、铁（石臼）捣茶和碾槽碾茶再筛罗。

（本图选自中国文物局编《惠世天工》，中国书店）

第二章 宋、辽、夏、金、元时代茶文化及茶画制作
Chapter Two Tea Culture and Tea-Related Painting Creation of the Song, Liao, Xia, Kin and Yuan Dynasties

101

图 045

北宋壁画·《夫妇对坐图》

　　元符二年（1099），墓葬壁画，高 92 厘米，宽 32 厘米。

　　夫妇二人对坐，应该是以茶为主题：酒瓶在桌子下面，处于密封状态。夫人后面侍女手捧圆盒，应该是茶末，男主人身后侍女所捧持的是渣斗——唐宋茶道中的滓方。

第二章　宋、辽、夏、金、元时代茶文化及茶画制作
Chapter Two Tea Culture and Tea-Related Painting Creation of the Song, Liao, Xia, Kin and Yuan Dynasties

103

图 046

北宋壁画·《夫妇对坐图》·洛阳市古墓博物馆藏品

　　靖康元年（1126），墓葬壁画。1983 年河南省新安县石寺乡李村北宋宋四郎墓出土。墓向南。位于墓室东部，为夫妇对坐场面。夫妇二人左右端坐于红色帷幔内，背后有屏风，面前一方桌，桌上摆放果品和茶、酒具。三侍者站立一旁：男主人身旁为男侍，女主人身旁为女侍，另一女侍立于墓主夫妇之间的后方。

图 047

北宋壁画·《夫妇对坐图》

　　墓葬壁画，高94厘米，宽87厘米。

　　1982年河南省登封市城南庄宋墓出土。墓向192°。位于墓室西壁。上绘横帐，方桌和两靠背椅。桌上放置二托盏，凳上袖手坐一妇人，头戴莲花冠，插步摇。站立两侍女，左女头梳高髻，捧粉红盘，内放二酒盏。右女头梳高髻，右手举执壶至胸前，似准备点茶。右椅空位，为男墓主之位。

第二章　宋、辽、夏、金、元时代茶文化及茶画制作
Chapter Two Tea Culture and Tea-Related Painting Creation of the Song, Liao, Xia, Kin and Yuan Dynasties

105

图 048

北宋壁画

　　河南省荥阳市槐西村 2008 年发现宋墓西壁壁画。高 86 厘米，宽 127 厘米。

　　夫妇对坐，桌前各有一套白瓷杯、红盏托，桌正后为一持执壶女子，左边为三个敲钹僧人。不同于一般家宴，又相别于法事活动。可能是以佛教音乐为助兴主题的夫妇宴居图。

图 049

北宋壁画

2011 年 11 月，郑州市文物考古研究院与登封市文物勘探队在河南省登封市唐庄乡进行文物调查时，发现西周时期遗址 1 处。唐宋时期窑址 1 处及汉代宋代墓葬 7 座。其 2 座宋代壁画墓较重要（《文物》2012 年第 9 期）。

上图为宋代 2 号墓西南壁与西北壁壁画。西北壁（右）为夫妇对坐品茶，他们后面站立仕女手捧点茶执壶，准备随时点茶。西南壁（左图）是三位仕女准备从茶酒间出门，奉送茶汤、茶点和食品。

北宋开封是全国点茶道的时尚之都，在登封一洛阳一带发现多个壁画墓，其品茶图不一而足，客观印证了北宋茶道炽盛的情景。

第二章 宋、辽、夏、金、元时代茶文化及茶画制作
Chapter Two Tea Culture and Tea-Related Painting Creation of the Song, Liao, Xia, Kin and Yuan Dynasties

107

图 050

北宋墓壁画·《仕女图》·中国国家博物馆藏品

　　墓葬壁画，河南省禹州白沙二号宋墓，高 80 厘米，宽 100 厘米。

　　图中间捧盘内为覆莲盖，红色盏，白瓷杯，这是标准的茶器组合。

图 051

宋墓壁画

　　墓葬壁画，与左图同属于一个墓室，与之对称地分布于墓室西南壁。图中，我们除看到白瓷杯和红色托盏外，似乎还可以判断：带碗注子就是点茶器，而且碗中没有温水，注子里是沸水。这里都是水果茶点，没有荤素菜肴。

第二章　宋、辽、夏、金、元时代茶文化及茶画制作
Chapter Two Tea Culture and Tea-Related Painting Creation of the Song, Liao, Xia, Kin and Yuan Dynasties

109

图 052

洛阳洛龙区关林庙宋代砖雕墓·砖雕画（《洛阳洛龙区关林庙宋代砖雕墓发掘简报》，《文物》2008 年第 1 期）

考古工作者推断为北宋晚期作品，点茶、奉茶的表现还是较为明确的。

图 053

宋代壁画·《奉茶恭候图》之一

　　山西长治市故漳村宋代砖雕墓墓室北壁西部壁龛两侧，画面左边为奉茶侍女。

第二章　宋、辽、夏、金、元时代茶文化及茶画制作
Chapter Two Tea Culture and Tea−Related Painting Creation of the Song, Liao, Xia, Kin and Yuan Dynasties

111

图 054

宋代壁画·《奉茶恭候图》之二

　　山西长治市故漳村宋代砖雕墓墓室东壁、西壁南部
壁龛两侧，右边奉茶，左边捧执壶。

图 055

宋代壁画·《奉果恭候》

　　山西长治市故漳村宋代砖雕墓墓室北壁西部壁龛两侧，水果、点心等被作为饮茶时的茶点来使用。

第二章　宋、辽、夏、金、元时代茶文化及茶画制作
Chapter Two Tea Culture and Tea-Related Painting Creation of the Song, Liao, Xia, Kin and Yuan Dynasties

113

图 056

北汉墓壁画·《备茶图》

　　北汉天会五年（961），墓葬壁画，高约 121 厘米，宽约 155 厘米。

　　1994 年山西省太原市第一热电厂北汉墓出土。壁画位于墓室西南柱间。穿着黑帽红衣的男子左手端持执壶，右手握防烫承托。图左白衣契丹少年捧白釉茶盏茶托。图右侧着圆领者双手捧一平沿束口白釉花瓶。桌上有各种水果。

图 057

北宋壁画·《医药图》·陕西考古研究院藏品

墓葬壁画，高 86 厘米，宽 145.5 厘米。

2009 年 3 月陕西省韩城市盘乐村 218 号墓出土，此壁画位于北壁下层。

画面正中画主人像。其他都是动态，除一女子奉茶，一女子执团扇，属于服务人员外，其他七名男子可以表明连贯的药房制药程序：端水盆（屏风后）—备药材—手举筛子—以铁臼碎药、罗药—依照典籍配置成药—奉药。我们以为这应该称为"制药房"，侍女为主人奉茶，穿蓝袍系红腰带的童子向主人奉药，捧书男子（画面最右）好像在看墓主人，或征询墓主人药剂的配伍剂量。其中典籍有宋太祖淳化三年（992）刊布的医典《太平圣慧方》。

第二章 宋、辽、夏、金、元时代茶文化及茶画制作
Chapter Two Tea Culture and Tea-Related Painting Creation of the Song, Liao, Xia, Kin and Yuan Dynasties

115

图 058

北宋墓壁画 · 《备茶图》

　　绍圣四年（1097），墓葬壁画，高 138 厘米，宽 79 厘米。

　　1999 年河南省登封市黑山沟村北宋李守贵墓出土。原址保存。

　　墓向 193°。位于墓室西南壁。上绘赭色幔帐、绿色横帐、绿、赭色组绶。帐下一方桌，桌饰云状牙条。桌上摆放盛桃子的四果盘和叠置的盏托等。桌左侧站立一妇人，头梳高髻，裹白色额帕，插步摇，戴耳环、手镯，外着褙子，内着抹胸，下着百褶裙，足着云头屦，右手捧茶罐（装茶末），左手持凤首茶匙向茶盏中添茶，表现点茶场景。桌后立一妇人，头梳包髻，戴白色额帕，外着褙子，下着百褶裙，面向桌左妇人抬手指点。左侧妇人身后置一屏风，屏风左侧地上置一火钳。

图 059

北宋墓壁画·王武子妻·《行孝图》

绍圣四年（1097），墓葬壁画，高 30 厘米，上宽 27 厘米，下宽 69 厘米。

1999 年河南省登封市黑山沟村北宋李守贵墓出土。原址保存。

墓向 193°。位于墓室西壁上方拱眼壁。门口竹帘高卷，内间左侧置灶台一，上放锅、碗。右侧立一黑色屏风，屏风前绘一卧榻，一老妇人头梳高髻，身穿褙子，内着团领衫、束抹胸，下着裤，盘腿坐于榻上。榻前一侍女，梳高髻，身穿褙子，下着百褶裙，双手捧托盏递与榻上老妇人。左侧一中年妇人，头戴包髻，一手执小刀，屈一足，表现"割肉奉亲"的场景。壁画左下角题记"王武子"，画面表现的是"王武子行孝，乳姑不怠"的故事。

第二章 宋、辽、夏、金、元时代茶文化及茶画制作
Chapter Two Tea Culture and Tea-Related Painting Creation of the Song, Liao, Xia, Kin and Yuan Dynasties

117

图 060

北宋墓壁画·《伎乐图》

绍圣四年（1097），墓葬壁画，高 136 厘米，宽 80 厘米。

1999 年河南省登封市黑山沟村北宋李守贵墓出土。原址保存。

墓向 193°。位于墓室西壁。上绘赭色幔帐、绿色横帐、绿、赭色组绶。中垂一同心结。画面有三个女子，均头梳包髻，外着褙子，内着抹胸。右边两人为女乐，右一人手捧笙吹奏，左一人手执拍板。身后一侍女，站在一方桌后，桌上一风炉。炉上置两个点茶的汤瓶，侍女手执其中一个准备点茶。

图 061

北宋墓壁画·《夫妇对坐图》

　　绍圣四年（1097），墓葬壁画，高 134 厘米，宽 77 厘米。
1999 年河南省登封市黑山沟村北宋李守贵墓出土。原址保存。
墓向 13°。位于墓室西北壁。上绘赭色幔帐、绿色横帐，青、
赭色组绶。帐下一方桌，桌上摆二茶盏、二果盘，桌旁对坐夫妇
二人。左侧女主人头梳高髻，裹额帕，身穿褙子，束红抹胸，下

着百褶裙，袖手坐于椅上。男主人头裹黑巾，着团领广
袖袍，腰束黑带，足着靴，袖手坐椅上。二人背后各立
一屏风，两屏风间立一侍女，梳包髻，戴耳环，双手端
注碗和注子。仕女左手捧茶执壶右手扶持防烫的承托碗。
过去人们往往将此壶、碗组合称为酒壶温碗，不一定都
是酒具。从这幅画看，壶与碗之间几无容纳热水的空间，
基本上刚好放下执壶而已，明显是点茶执壶。

第二章 宋、辽、夏、金、元时代茶文化及茶画制作
Chapter Two Tea Culture and Tea-Related Painting Creation of the Song, Liao, Xia, Kin and Yuan Dynasties

119

图 062

北宋墓壁画·《仕女图》

　　原画高 155 厘米，宽 210 厘米。

　　河南济源市东石露头村出土，墓葬壁画，后站妇女双手动作不清晰。图左侧（右前方）妇女左手端

白釉斗笠形茶盏、黑色釉茶托，右手可能提执壶；右侧妇女双手端盏托。

图 063

辽墓壁画

　　2000年辽宁省朝阳市龙城区召都巴镇出土。墓向190度。分别为墓室东壁前半部和西壁前半部。两壁都有茶盏茶托。东壁画中有侈口酒尊，西壁则是带盖茶叶罐。图中人物施叉手礼。（《中国出土壁画全集》（8）科学出版社2012年1月第一版。）

图 064

金代墓壁画·《点茶图》

　　北京，墓葬壁画，皇统三年（1143）墓。点茶的执壶和茶托茶盏山都很标准，茶盏托与陕西蓝田吕氏家族墓的几乎一模一样。点茶的是汉人，做助手的是年轻的契丹人。值得注意的是桌子上的盝顶带两层方圈足盒，很可能是装放茶末的器具。后面一人半举带纱网罩的可能是调料盘。

第二章　宋、辽、夏、金、元时代茶文化及茶画制作
Chapter Two Tea Culture and Tea-Related Painting Creation of the Song, Liao, Xia, Kin and Yuan Dynasties

121

图 065

北京大兴辽墓 M1 北壁墓门东侧壁画和 M1 南壁墓门东侧壁画

　　左图二人，一人拿白色披巾，另一人端白釉茶碗、黑釉茶托。右图为一妇女端捧长颈酒壶。

图 066

辽 · 彩绘木雕说法会群像 · 西安私人收藏

　　可名之《佛说法图》。与白描线画、壁画、流传密教佛画和唐卡不同，这组辽代木刻的佛前床案上是一组茶壶和直筒茶杯、果盒。

第二章　宋、辽、夏、金、元时代茶文化及茶画制作
Chapter Two Tea Culture and Tea-Related Painting Creation of the Song, Liao, Xia, Kin and Yuan Dynasties

123

图 067

山西大同东风里辽代壁画墓东壁壁画

M1（一号墓）墓室东壁壁画图中有五位侍者。周围有莲花柳枝、马鞍、牡丹、竹子、鹿、鹤、龟、散落的羊骨节形（或马鞍形）金饼、金钱；有花瓶。各种吉祥寿康元素的组合在画面右侧。五人从左到右分别拿乐器、石榴等水果盘，端有两个酒盅的圆盘、花束、细长颈酒壶和六瓣弧曲形浅腹温酒碗。画面左下方为三只茶执壶放在五足铜炭炉上正烧水、一大一小的酒瓶、圆形套装食盒。从这些元素可断定：以契丹辽人为主的茶酒宴会即将开始。点茶执壶直接在炉火上烧水，酒壶的瓶颈很细，带盖，为我们判断茶酒器之不同提供了非常鲜明的图像资料。

图 068

M1（一号墓）墓室西北壁《虎形灵兽图》

图 069

M1（一号墓）墓室西北壁《钓鱼图》

第二章 宋、辽、夏、金、元时代茶文化及茶画制作
Chapter Two Tea Culture and Tea-Related Painting Creation of the Song, Liao, Xia, Kin and Yuan Dynasties

125

图 070

M1 墓室西北壁《宴饮图》

　　M1 西北壁《宴饮图》是整个墓室壁画的中心。男人双手捧杯，仿佛表示"先干为敬"，端的是白瓷杯；夫人端坐，手有披巾遮掩。桌上有小酒壶（相当现代的分酒器）、酒盅和酒盅座，仿佛也是白色杯子与黑色红色的托子组合。按比例看是较小的酒盅。这仿佛告诉我们酒宴刚开始，茶还没有上。

图 071

M1（一号墓）发掘全景

图 072

东壁壁画

　　M1 东壁壁画（《文物》，2015 年第 3 期），可名为《斟酒图》。图中七人各拿东西，最中间两人，一人捧持酒瓶正在倒酒，另一人正端酒盅酒台子盛酒。四直腿、带四格护栏的方形桌子上，有酒壶套在温碗中，一个大，另一套较小（似为茶执壶），两只倒扣的圈足碗、两只白色茶杯（茶杯明显比酒盅大）、带勺的深腹汤品。这幅壁画明确告诉我们：1. 茶杯大、酒盅小。2. 酒直接倒进酒盅，并不一定要加温。

第二章　宋、辽、夏、金、元时代茶文化及茶画制作
Chapter Two Tea Culture and Tea—Related Painting Creation of the Song, Liao, Xia, Kin and Yuan Dynasties

127

图 073

河北宣化辽金壁画墓东壁壁画（局部）

　　这是宣化新发现的壁画，可名为《点茶图》。

图 074

河北宣化辽金壁画墓南壁壁画（局部）

与图 073 为组合壁画，可名为《男侍进侍图》，最前面红衣侍者双手捧白釉瓷渣斗，即茶事中的滓方。

图 075

河北宣化辽金壁画墓南壁壁画

　　宣化新发现壁画。男女侍者恭立图。(《文物》2013 年第 12 期)

图 076

河北宣化辽金壁画墓西壁壁画

　　宣化新发现壁画（《文物》2013 年第 12 期）。可命名为《茶事间》或《备茶间》。此图明确表明茶执壶与酒壶有明显差异：茶执壶高挑，壶口向上呈喇叭口外撇，流口较直。而酒壶一般是细颈直口。三套白釉茶盏配红釉茶托（红漆茶托）成为宋辽金的一般组合。另有茶点盘、茶末红罐（在画面前方靠近执壶茶托）。红布包囊里面可能是茶末罐。令人深思的是，从两位一长（左侧红衣捧茶盘）一幼站立恭候、表情庄重神情、有衣冠架及红蓝衣服等看，辽代是否已经出现以事茶为主要技能的职业女子？

第二章　宋、辽、夏、金、元时代茶文化及茶画制作
Chapter Two Tea Culture and Tea−Related Painting Creation of the Song, Liao, Xia, Kin and Yuan Dynasties

131

图 077

辽·《备茶图》（摹本）

　　墓葬壁画，高约 126 厘米，宽约 246 厘米。

　　1985 年山西省大同市纸箱厂辽墓出土。已残毁。

　　墓向 187°。位于墓室东南壁。中间一方桌，摆放一篓笥、盏把、盏。左侧一侍女端一托盏，一侍女跪坐，右手持一物，面前似有一火盆，似在备汤。其右侧站立二侍女，一侍女端一大碗，一侍女捧一小盘。三站立女侍从左到右分别是茶盏茶托、白釉碗、小碟。跪坐侍女可能正在候汤，似乎因火焰大，而用手遮面。桌上有一木盒子，另有一茶盏、茶托分别放置，似是用过等待清洗归整。

图 078

辽·《备茶图》

墓葬壁画，高约 72 厘米，宽约 110 厘米。

1991 年山西省朔州市市政府工地辽墓出土。已残毁。

墓向 95°。位于墓室南壁。上方为额枋，左侧五名侍仆围绕在桌旁忙碌，桌前一人在风炉旁煎汤，桌后二人分别持渣斗、盏托，桌上置托盏、茶碾，其左侧二人奉递方盒、盏托。右侧两名侍女正在点茶，旁边还站立一名髡发的男童、一名梳髻捧持茶盏的女童。这是典型的《备茶图》：后排，从左到右分别是绿衣男子焦急眺望；绿衣女子清洗白釉茶盏；两男侍配合点茶，一提执壶，一捧盏托；蓝衣侍女正接绿衣侍女从客厅送进来的酒盅；穿蓝衣圆领服契丹青年手捧白釉渣斗；绿衣青年侍女手捧方盒；戴幞头蓝衣男子正用火箸添加木炭，茶执壶在火炉中煮水。这是"茶酒间进中"，即酒退茶进的更换之间。第一个绿衣男子可能刚从客厅跑到后厨茶水间催促大家：快上茶！

第二章 宋、辽、夏、金、元时代茶文化及茶画制作
Chapter Two Tea Culture and Tea-Related Painting Creation of the Song, Liao, Xia, Kin and Yuan Dynasties

133

图 079

辽·《备茶图》

天庆六年（1116），墓葬壁画，人物高约119厘米。

1974年河北省宣化下八里1号张世卿墓出土。原址保存。墓向207°。位于后室东壁。朱红色的桌子上摆放着托盏，典型的白盏黑釉茶托、汤瓶、香炉、香合盒和黄色函盒等，在桌子的外侧还放有《常清净经》和《金刚般若经》。两个人物头戴黑色交脚幞头，右侧人物着皂色长袍，双手扶大盘口汤瓶。左侧人物着赭色长袍，回首顾盼，右手扬起，似和左侧人物交谈。二人身后为直棂窗一扇。

图 080

辽·《备酒图》

天庆六年（1116），墓葬壁画，人物高约 114 厘米。

1974 年河北省宣化下八里 1 号张世卿墓出土。原址保存。墓向 207°。位于后室南壁西侧。画中绘红色方桌，桌上有椭圆长盘、酒盏、盘口瓶、执壶等，桌后站立的两个男子，左侧的身穿深褐色圆领长袍，双手捧一黑色盘，盘内有酒杯一对；右侧人物身着皂色圆领长袍，腰束带，手捧白色执壶与成套的温碗。桌前方绘木架，架上插放三个带盖梅瓶，上绘朱色"张记"印封帖。桌旁绘灯檠。

桌前端三瓶酒，一瓶启封，两瓶未开启，外黑内白色瓷盘中的三个杯子可能是酒盅。两组倒立的斗笠形茶盏，可能因茶品不同，而用不同茶盏。

先进酒，还没有开始点茶。

第二章 宋、辽、夏、金、元时代茶文化及茶画制作
Chapter Two Tea Culture and Tea-Related Painting Creation of the Song, Liao, Xia, Kin and Yuan Dynasties

135

图 081

河北省宣化辽墓 M1 后室西壁《点茶图》

天庆六年（1116），墓葬壁画，人物高约102厘米。1974年河北省宣化下八里1号张世卿墓出土。原址保存。

墓向207°。位于后室西壁。朱红色的桌子上摆放着茶盏、瓷盆、漆盒等，桌下有一圆形炭盆，炭火上放一汤瓶。桌旁是正在备茶的两个人物，左侧黄袍的老者，左手托黑托白盏的托盏，右手持茶匙拨动盏中的茶末。右侧的人物左手扶桌面，右手执黄色汤瓶，准备为老者点茶。

河北省宣化辽墓 M1 后室西壁《点茶图》（局部）

很明显，宋辽时期白釉盏配黑釉茶托，这是与宋代斗茶崇尚汤色白、泡沫耐久有关，尚需再研究。

第二章 宋、辽、夏、金、元时代茶文化及茶画制作
Chapter Two Tea Culture and Tea-Related Painting Creation of the Song, Liao, Xia, Kin and Yuan Dynasties

137

图 082

宣化辽墓 M7 前室东壁《茶酒间》

　　这是一间很特殊的房间：有文房四宝、书籍、木匣，有茶盏、酒杯，有酒瓶。深色波罗子放桌上，白酒瓶与白釉温碗中间深色套盒，与法门寺地宫出土波罗子很相似，极有可能是贮茶器。挂在屋顶吊钩上的竹篮，应是时蔬、糕点或水果及其他食品。画中人物八人：四个小孩躲在桌案、组套食式柜后戏耍偷窥大人取干果或水果；一年轻女子手捻花蕾；一髡发青年跪地为梯，让一妇女取高吊的篮子里的东西；另有一青年用袍服装盛取出来的东西（很像菱角）。

　　实际这就是备茶备酒的"茶酒间"。因为是现场用茶碨轴碾磨炙青饼茶，因此不见茶末罐，但见到茶刷子、擦布和托盘。有未打开封盖的酒瓶，有笔墨纸砚。桌子上的执壶和套碗告诉人们：茶执壶也套碗，但这只碗的用途仅仅是防止烫手，因为茶执壶直接在炉火上加热。陶碗刚好容下执壶，也就没有多少空间盛贮热水以温酒。也就是说，辽宋的执壶套碗一起使用，并不都是酒具。

图 083

宣化辽墓 M7 后室东壁《奉茶图》

第二章 宋、辽、夏、金、元时代茶文化及茶画制作
Chapter Two Tea Culture and Tea-Related Painting Creation of the Song, Liao, Xia, Kin and Yuan Dynasties

139

图 084

宣化辽墓 M10 前室东壁《备茶图》

　　碾茶者旁边的漆盘里有一个直筒杯，是用来贮存末茶的。方
桌上有茶则，一茶筅，细颈、瓜棱执壶及其竹编提篮。与宋·刘
松年《斗茶图》中的器具基本相同。

宣化辽墓 M10 前室东壁《备茶图》·碾茶（局部）　　　　宣化辽墓 M10 前室东壁《备茶图》·吹火（局部）

第二章　宋、辽、夏、金、元时代茶文化及茶画制作
Chapter Two Tea Culture and Tea-Related Painting Creation of the Song, Liao, Xia, Kin and Yuan Dynasties

141

图 085

宣化辽墓 M6 前室东壁《备茶图》

　　这幅图告诉我们，陆羽《茶经》在传播过程中，有一个民族化、地方化、阶层化的改造过程。图右上角是一个小男孩，趴在盝顶食式柜上打瞌睡，从打开的三层抽屉看，这相当于都篮的功能；出现了《茶经》与《大观茶论》中所没有的茶具：锯子。茶则、夹子（镊子）、勺子、茶笕与唐宋的一致。与法门寺皇帝使用碨轴相比，底座升高，加大轮饼直径，加了前后使力的推架，可以一个手碾茶，不用放在桌子、床几上，小孩坐在地上就可以碾茶。折腰六棱带流高执手细颈直口茶执壶，腹大口小，很易于与酒壶混淆。

图 086

宣化辽墓 M5 后室东南壁《点茶图》

　　穿深色圆领衫服的契丹人用带敞口、高足碗的执壶正在倒沸水。这表明，碗是用来降温、防烫。点茶过程与带汤勺盆缶同处于一个画面，喝茶与汤肴同时饮用。

第二章 宋、辽、夏、金、元时代茶文化及茶画制作
Chapter Two Tea Culture and Tea-Related Painting Creation of the Song, Liao, Xia, Kin and Yuan Dynasties

143

图 087

宣化辽墓 M5 后室西南壁《备茶图》

　　五足铁炭炉在北方常见。执壶直接在火炉上煮水的方法延续很久，明代绘画中尚有同样的
形式。这幅图很容易使人猜想，先在盆缶中点茶，再舀进茶盏。但联系上一图，我们还是认为
单盏单点可能更普遍、正规。

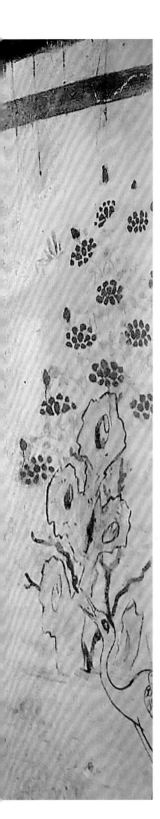

图 088

宣化辽墓 M2 墓室南壁东侧《奉茶图》

第二章 宋、辽、夏、金、元时代茶文化及茶画制作
Chapter Two Tea Culture and Tea-Related Painting Creation of the Song, Liao, Xia, Kin and Yuan Dynasties

145

图 089

宣化辽墓 M2 墓室东南壁东侧《备经图》

一人袖手旁观、一人持渣斗，桌上有经函、插花。

图 090

宣化二号墓壁画

　　这是宣化近年新出土的辽墓壁画。应该注意两点：茶执壶与套碗很紧密，几无缝隙，就排除了温酒器可能；客人多，连茶托都简化掉了，直接用小碟子盛端茶盏。

第二章　宋、辽、夏、金、元时代茶文化及茶画制作
Chapter Two Tea Culture and Tea-Related Painting Creation of the Song, Liao, Xia, Kin and Yuan Dynasties

147

图 091

辽·《备茶图》

天庆元年（1111），墓葬壁画，高 142 厘米，宽 131 厘米。

1990 年河北省宣化下八里 4 号韩师训墓出土。原址保存。

墓向 178°。位于后室东南壁。描绘的是一添灯侍女和三个围在案前备点茶的侍女。案上放执壶、茶盏、托子等，四周有朱栏，柱间为云板，栏杆中开两个口，以便取物。与以前见过的茶执壶、茶托盏看，此图中的执壶与盏托比例有变化，执壶小了，茶盏大了。

图 092

辽·《宴乐图》（品茶赏乐）

天庆元年（1111），墓葬壁画，高 142 厘米，宽 14 厘米。

1990 年河北省宣化下八里辽 4 号韩师训墓出土。原址保存。

墓向 178°。位于后室西南壁。画面可分听琴者和演奏伴唱者两部分。画面右侧在放着果盒和漆盘的红色方桌前，一妇人白巾包髻，身穿短袄长裙，坐在圆形座墩上，手捧托盏，在专心地欣赏乐舞。妇人的身旁是一双手持盘的老者。画面左侧有三人，舞旋者双手击掌，双腿弓曲，随着琴声的节奏在舞蹈。旁边两人，一长者在弹琴，另一长者，身子侧向女主人，而回望弹琴者，双手击掌。乐舞者身后有长桌，上置茶酒具，身前为带盖梅瓶并架。高腿方桌上的茶执壶颇有特点：西瓜形、无颈、短直流。

第二章 宋、辽、夏、金、元时代茶文化及茶画制作
Chapter Two Tea Culture and Tea-Related Painting Creation of the Song, Liao, Xia, Kin and Yuan Dynasties

149

图 093

辽宁法库县叶茂台辽肖义墓墓门旁西侧壁画《献茶图》

　　墓葬壁画，高 171 厘米，宽 91.5 厘米。

　　主体画面为两人合作点茶。桌下另有一人，蹲在火盆旁，手拿火箸，在拨弄炭火，做温酒状。炭火上座着一尊长颈壶和有扣盖的小罐，茶盏、托显得瘦高。

图 094

辽宁法库县叶茂台辽肖义墓墓门东侧壁画

　　左边一人捧方盘，盘中放两碗；右边一人双手捧持大碗，碗内一长柄勺。桌上有酒盅数枚，桌下两瓶封口酒坛。

　　辽萧义（1038—1111）于1112年葬于辽川之右圣迹山阳衬（祔）先茔也，其夫人耶律氏为陈国夫人。该墓发掘于1976年4月13日至9月7日。我们觉得前一幅为备茶献茶，后一幅可名为汤肴图。（温丽和《辽宁法库县叶茂台辽肖义墓》，《考古》1989年第4期）

第二章 宋、辽、夏、金、元时代茶文化及茶画制作
Chapter Two Tea Culture and Tea-Related Painting Creation of the Song, Liao, Xia, Kin and Yuan Dynasties

151

图 095

金 · 《点茶图》

　　墓葬壁画，内蒙古滴水洞金墓壁画。这也是典型的两人合作点茶
图，左侧一人奉送茶酒菜肴。

图 096

辽·《厅堂图》

天赞二年（923），墓葬壁画，高 208 厘米，宽 270 厘米。

1994 年内蒙古阿鲁科尔沁旗东沙布日台乡宝山村 1 号墓出土。原址保存。

墓向正南。位于石室北壁。主要表现厅堂的布置。壁上挂着弓箭、弓囊和剑，地面铺有地毯，上摆桌椅，桌上有杯、盘、碗、盏。桌、椅边角装饰贴金箔。我们可以把这里看作蒙古人的茶室。

第二章 宋、辽、夏、金、元时代茶文化及茶画制作
Chapter Two Tea Culture and Tea-Related Painting Creation of the Song, Liao, Xia, Kin and Yuan Dynasties

153

图 097

辽·《奉茶图》

大安三年（1087），墓葬壁画，人物高 175 厘米。

1997 年巴林右旗索布日嘎苏木辽庆陵陪葬墓耶律弘世墓出土，原址保存。

墓向 160 度。位于西耳室甬道北侧。汉装男侍双手端红托白盏茶器。这位魁梧挺拔的汉装男子恭敬地捧着茶托茶盏，令人们可以想象茶的珍贵与饮茶者非同一般的地位。

图 098

辽·《点茶图》

墓葬壁画，高 120 厘米，宽 120 厘米。

1990 年内蒙古巴林左旗查干哈达苏木阿鲁召嘎查滴水壶辽代墓出土。壁画原址保存，摹本存巴林左旗博物馆。

其左侧头裹黑巾者所提执壶带圈足，壶口为敞口长细颈，提梁偏向一侧与另一侧曲流相对。右侧侍者双手端持黑托白釉敞口深腹盏，估计实物口径超过 10 厘米。这组配套茶器在我们所见壁画中并不多见。但茶盏、茶托异色搭配同样在宋辽金元墓壁画和宋代、明代国画中较为普遍。

第二章 宋、辽、夏、金、元时代茶文化及茶画制作
Chapter Two Tea Culture and Tea-Related Painting Creation of the Song, Liao, Xia, Kin and Yuan Dynasties

155

图 099

辽·《品茶图》

　　墓葬壁画,高136厘米,宽136厘米。

　　1995年内蒙古敖汉旗四家子镇闫杖子村北羊山1号墓出土,现存敖汉旗博物馆。

　　正面三人戴交脚幞头,身穿圆领窄袖长袍。两侧为契丹侍仆,髡发。桌前一男孩酣睡,一女孩在烧火煮茶执壶之水。这幅壁画告诉人们:其上还可放置大一点的瓷碗,放置各类食品;而最右侧男子左手端着的白釉色直口深腹杯,从大小看,应是茶杯。

图 100

辽·《奉茶图》

1991 年内蒙古敖汉旗南塔子乡城兴太村下湾子 1 号墓出土,现存敖汉旗博物馆。

墓向 150°。壁画位于墓室东壁。图中男主人头裹黑巾,深褐色长袍,双脚穿白袜歇踏于方木墩之上。似为夫妇对坐图。男主人跟前的红色茶托和青色茶盏相搭配。茶盏为多曲葵口带高圈足,后边穿红色衣服侍女双手端持红色温碗青白釉斜直流直口短颈执壶。这幅图的一个重要参考意义在于它告诉人们,带温碗的执壶也是点茶器。而人们以往常将这种搭配解读为温酒器。此外,桌子上有水果、茶点。

第二章 宋、辽、夏、金、元时代茶文化及茶画制作
Chapter Two Tea Culture and Tea−Related Painting Creation of the Song, Liao, Xia, Kin and Yuan Dynasties

157

图 101

金·《奉茶散乐图》

　　正隆六年（1161），墓葬壁画，墓向180°，位于徐龟墓室西壁。原图被定名为《奉酒散乐图》，比较准确。画面最右侧妇女手端方盘，盘内茶盏为六曲葵口斜腹带玉璧足。与黑漆方桌上带红色托子白釉茶盏差别较大。另外桌上还有敞口斜腹小酒盅。带温碗方形斜直流执壶也可能就是茶器执壶，方形壶实物不多见。茶点为多种水果。桌下为两个黑色酒坛。用带温碗执壶里的沸水在瓷匜（外钴蓝釉，内白釉）点茶，然后倒入四瓣海棠花口式盆钵（见图最左侧侍女）。桌上有两个钴蓝色瓷匜，左侧侍女手捧其中一个倾茶汤（白色），一个在桌前正中，可见白色茶汤。

　　至少用两个以上的瓷匜点茶，然后倒入盆钵，以供多人同时品茶。

　　桌上有瓜棱形葵口瓷茶盏配红釉瓷托子，有斜腹白釉酒盅——其口径几乎只有茶盏一半。

图 102

大同金徐龟墓之东壁北侧之《奉茶图》

　　墓葬壁画，1990年发现。图右侧有一执壶，但人物形貌不清，只能命名为《奉茶图》。（《考古》2004年第9期）

第二章 宋、辽、夏、金、元时代茶文化及茶画制作
Chapter Two Tea Culture and Tea-Related Painting Creation of the Song, Liao, Xia, Kin and Yuan Dynasties

159

图 103

金·家族墓地五号墓

　　山西省汾阳市金代家族墓地五号墓，壁画上有牌匾"茶酒位"。画中黑衣人左手
端盏——标准的外黑内白斗笠形，右手执茶筅，点茶。"茶酒位"，应解读为茶酒间——
准备茶、酒的专门空间。点茶器不是唐宋常用的斗笠盏，而是内白外黑的瓷匜。

图 104

金·《男侍图》·现存于林州市博物馆

　　墓葬壁画。2006 年河南省林州市桂林镇三井村金墓出土。墓向南。
位于墓室北壁上层，为一男侍站在一方桌子边，桌上有托盏一副及果盘、
茶匙，男侍头裹黑巾，身穿白色团领窄袖短衫，下穿白色长裤，正往桌
子上摆放食物。

第二章 宋、辽、夏、金、元时代茶文化及茶画制作
Chapter Two Tea Culture and Tea-Related Painting Creation of the Song, Liao, Xia, Kin and Yuan Dynasties

161

图 105

金·《奉茶进酒图》（局部一）

大定九年（1169），墓葬壁画，高约 140 厘米，宽约 103 厘米。

2007 年山西省陵川县附城镇玉泉村金墓出土。原址保存。

墓向 180°。位于墓室东壁南侧。左前一男仆，恭敬站立，端一劝盘三酒盏；右侧下方一男仆，左手撑膝，右手执一壶伸向塘炉。在一围幔的高桌后面，站立着两男仆，左者捧着带温碗的注壶，回顾右方；右者左手握着长颈瓶、右手以抹巾托着瓶底，正在与前者说话。画面最前方一人正用茶执壶煮水，酒壶茶壶差异明星。

图 106

金 · 《奉茶进酒图》（局部二）

　　大定九年（1169），墓葬壁画，高约 140 厘米，宽约 43 厘米。

　　2007 年山西省陵川县附城镇玉泉村金墓出土。原址保存。墓向 180°。位于墓室南甬道口东侧。一男仆，侧背身站立在一摆放着盏托的高桌之前，正将一副茶盏托从桌案右侧一橱柜里取出，斗笠形茶盏重叠倒扣在方形橱柜（都篮）之中；高桌前下方一方形束腰矮桌，放置两个耸肩的圆酒坛。这个墓室的东壁南壁壁画告诉我们：1.先奉酒，后进茶；2.此茶酒间非常专业，上储存茶器，下储存酒瓶。可能是高门富户的后厨标准设置。

第二章 宋、辽、夏、金、元时代茶文化及茶画制作
Chapter Two Tea Culture and Tea-Related Painting Creation of the Song, Liao, Xia, Kin and Yuan Dynasties

163

图 107

金一元 · 《厅堂图》

金元墓葬壁画，高约 76 厘米，宽 120 厘米。

1986 年河南省禹州市坡街村壁画墓出土。原址保存。

墓向 190° 。位于墓室东南壁。画面右侧为一屏风。屏心绘有山水人物；屏风之上绘有悬幔，屏风后一青年侍女手推悬幔探身欲进。画面左侧绘一桌，在长方桌面上置有白色的盘、酒瓶、酒壶、花口碟和茶托等物。近桌处立一侍女，双手捧一圆盘，内盛黄色果品。

图 108

金·《侍茶图》

大定九年（1169），墓葬壁画，高约 140 厘米，宽约 103 厘米。

2007 年山西省陵川县附城镇玉泉村金墓出土。原址保存。

墓向 180°。位于墓室西壁南侧。左侧一男仆，举一笼盖茶盏的托盘，向右侧走来。右侧一垂幔高桌，摆有一摞三件的盏托；桌案后面站立二男仆。左者，侧首右顾，其左手斜拿一盏托，右手向右指点；右者，平端一盏托，面向前者，作聆听状。这表示客主要开始品茶了！

第二章 宋、辽、夏、金、元时代茶文化及茶画制作
Chapter Two Tea Culture and Tea-Related Painting Creation of the Song, Liao, Xia, Kin and Yuan Dynasties

165

图 109

金·《庖厨图》

大定九年（1169），墓葬壁画，高约 140 厘米，宽约 83 厘米。

2007 年山西省陵川县附城镇玉泉村金墓出土。原址保存。

墓向 180°。位于墓室南甬道口西侧。在两张高桌后面，有三名男仆。左侧一仆，俯身面向砧板，持长刀以分切。右侧站立二人，左者端一案三盏，侧身回顾；右者手拿一长柄圆勺，左手端一盏，面向前者说话。他们的身后还有两张高桌，摆放着食盒及大盆、小碟等。他们可能准备面条吧。而三个直筒杯无疑是茶杯。似乎饭后一杯茶，成为一种搭配。从其专业的制服看，他们有点像是一个专业的炊事点茶服务队。

图 110

金·《"香积厨"图》

明昌六年（1195），墓葬壁画，高约 148 厘米，宽约 70 厘米。

2008 年山西省汾阳市东龙观村金代家族墓地 5 号墓出土。原址保存。

墓向 190°。位于墓室西北壁，壁画上部有墨书牌匾"香积厨"三个字，下部绘有两个侍女，一人手中捧着一箩刚出锅的包子，另一个手中持一个托盘，盘中放置三个碗；两位侍女相互顾盼、照应，趋向墓后壁的方向。

第二章 宋、辽、夏、金、元时代茶文化及茶画制作
Chapter Two Tea Culture and Tea-Related Painting Creation of the Song, Liao, Xia, Kin and Yuan Dynasties

167

图 111

金·东北柱间壁全景·现存于山西博物院

墓葬壁画，高约 98 厘米，宽约 124 厘米。

2007 年山西省繁峙县杏园乡南关村金墓出土。

墓向 180°。位于墓室东北柱间壁。中间砌筑一涂红框的白色菱花窗，窗户上、下有各式珍宝奇物。其右侧一名文官端坐于高桌之后，头戴展脚幞头，身着圆领窄袖长袍，桌上有账簿一本。文官身后竖立一面素面屏风、两竿修竹。画面左侧站立五名侍女、一髡发男侍，分别捧持包袱、长盘、渣斗、珊瑚、铜镜等物。

图 112

金·《内宅图》

墓葬壁画，高约 78 厘米，宽约 112 厘米。

1994 年山西省平定县城关镇西关村 1 号金墓出土。原址保存。

墓向 176°。位于墓室北面柱间壁。垂帐高悬，笼盖一席床榻，当间挂一绣球。榻前两侧各有一名侍从。左侧一女包髻，披帛，身着褙子，拱手站立；右侧一男，髡发，身着圆领窄袖长袍，持巾侍立。如果是洞房图，他们就是主要服务对象。

第二章 宋、辽、夏、金、元时代茶文化及茶画制作
Chapter Two Tea Culture and Tea-Related Painting Creation of the Song, Liao, Xia, Kin and Yuan Dynasties

169

图 113

金·《出行图》

墓葬壁画，高约 78 厘米，宽约 112 厘米。

1994 年山西省平定县城关镇西关村 1 号金墓出土，原址保存。

墓向 176°。位于墓室西北柱间壁。右侧一男侍站立，头戴展脚幞头，身着圆领窄袖长袍，双手持骨朵导引，回首左顾；第二人头戴短脚幞头，身着圆领窄袖长袍，双手托一包袱（食盒）；第三人头戴短脚幞头，身着圆领窄袖长袍，抱一束颈带盖瓶；随后一名老翁，头戴无脚幞头，身着圆领窄袖黄色长袍，左手提一茶笼，右手拎一茶壶。除第一人外，后边三人所带包袱、带盖瓶、提篮、包裹等，似乎与茶酒有关。

图 114

金·《奉酒图》

墓葬壁画，高约 78 厘米，宽约 112 厘米。

1994 年山西省平定县城关镇西关村 1 号金墓出土，原址保存。

墓向 176°。位于墓室西面柱间壁。右侧一人头戴短脚幞头，身着圆领窄袖长袍，回首，招呼另外三人向前行。随行两人提一斗箱，内装满酒坛，旁边站立一童仆，手捧托盘，内似装满小酒瓶。这两幅壁画都是四人图，前边都有一个官员引导，是举行重要仪式的一部分。

第二章 宋、辽、夏、金、元时代茶文化及茶画制作
Chapter Two Tea Culture and Tea-Related Painting Creation of the Song, Liao, Xia, Kin and Yuan Dynasties

171

图 115

金·《朱俊少氏对坐图》

明昌四年（1193），墓葬壁画，高 36 厘米，宽 70.05 厘米。

2008 年陕西省甘泉县城关镇袁庄村金墓出土。原址保存。

墓向南。位于东壁中。画面正中为一黑色高足方桌，桌上有三碟、两盏。方桌后有山水屏风画。

桌右侧坐一老年男子，身后上方题写"朱俊"二字，应为老者姓名。朱俊身后侍立一中年男子，头部

上方墨题"男朱孜"。方桌左侧，坐一老年妇人，头部上方墨题"少氏"。老妇人身后侍立一年轻妇人，头部上题书"高氏"。此画面右侧边有一行墨书题记："明昌四年十一月初一日工毕"。从题记推测，此画面是墓主人朱俊与少氏的对坐饮宴图。从形制看，有可能女喝茶，男饮酒。

第二章 宋、辽、夏、金、元时代茶文化及茶画制作
Chapter Two Tea Culture and Tea-Related Painting Creation of the Song, Liao, Xia, Kin and Yuan Dynasties

173

图 116

陕西甘泉金代壁画墓四号墓东壁《听琴图》

图 117

陕西甘泉金代壁画墓四号墓北壁《对弈图》

图 118
陕西甘泉金代壁画墓四号墓北壁《背诵诗文图》（似应定名为《握笔作书图》）

图 119
陕西甘泉金代壁画墓四号墓北壁《赏画图》

第二章 宋、辽、夏、金、元时代茶文化及茶画制作
Chapter Two Tea Culture and Tea-Related Painting Creation of the Song, Liao, Xia, Kin and Yuan Dynasties

175

图 120

陕西甘泉金代壁画墓四号墓西壁《策马行旅图》

　　M2、M3、M4 三墓中壁画内容丰富，题材新颖，既有河南、山西、甘肃等地宋金时期墓葬壁画中常见的宴饮图、备宴图和孝行故事，又有山水、木石、群鹤、孔雀、荷塘、天鹅、人物出行等上述地区较为少见的自然题材，而且群鹤图、山水图在三座墓中多次出现，反映了甘泉地区金代墓葬壁画选材方面的地方特点。特别是 M4 的 11 幅壁画中，没有宴饮、备宴、孝行故事等常见的与奉孝相关的内容，却出现了体现个人情趣的听琴图、弈棋图、诵书图和赏画图，即所谓"文人四艺"——琴、棋、书、画；东侧阑额上亦绘有兰花、梅枝、菊花、竹子图案，即"花中四君子"——梅、兰、竹、菊，题材内容较特殊，目前尚未见到类似的报道。由此可见，后世流行的"文人四艺"和"花中四君子"的内容构成及组合在金代中期的民间已经定型。这种突出表现墓主生前情趣修养的做法，与 M1 壁画中浓厚的奉孝氛围形成了鲜明对比。我们推测，M4 的墓主应是当地有一定文化修养的富有乡绅。（王勇刚、刘立平、高强等《陕西甘泉金代壁画墓》，《文物》2009 年第 7 期）宋金时期，甘泉为延安府所辖，属于延安路。据近年调查，延安地区中部分布有许多宋金时期的仿木结构雕砖墓和壁画墓，但由于基本未经科学发掘，其整体面貌目前尚不很清楚。甘泉袁庄金代纪年壁画墓的发现，为研究金代建筑、服饰、社会生活、丧葬习俗等提供了珍贵资料，更为这一地区宋金时期同类墓葬的分期研究提供了一个重要标尺。如果我们统筹考虑几个建造于金大定二十九年（1189）至明昌四年（1193）的朱氏四座墓葬，可以感到茶香与文人四艺的结合的史实。这里既非传统文化厚重的关中腹地，又不是富庶繁华的江南茶乡，但我们分明从壁画中感到浓郁芬芳的茶的真香和琴棋书画的美妙氤氲。这从一个侧面折射出宋金时代中华茶道文化极为强大的文化影响力和融合力。唐宋茶道与琴棋书画互为融摄。甘泉金墓表明茶在女真金国也有普遍深入发展。

图 121

山西屯留宋村金代壁画墓墓室东壁

图 122

山西屯留宋村金代壁画墓北壁左侧《夫妻对坐图》

　　图中一对老年夫妇，桌案上各有一套茶盏、茶托。

第二章　宋、辽、夏、金、元时代茶文化及茶画制作
Chapter Two Tea Culture and Tea-Related Painting Creation of the Song, Liao, Xia, Kin and Yuan Dynasties

177

图 123

山西屯留宋村金代壁画墓墓室北壁

图 124

山西屯留宋村金代壁画墓东壁墓主人夫人像

侍女双手捧持托盘里应是白瓷斗笠形茶盏。

第二章 宋、辽、夏、金、元时代茶文化及茶画制作
Chapter Two Tea Culture and Tea-Related Painting Creation of the Song, Liao, Xia, Kin and Yuan Dynasties

179

图 125

山西屯留宋村金代壁画墓东壁北壁墓主人像

后站立仆人所捧持之物应为渣斗。

图 126

山西屯留宋村金代壁画墓北壁东侧壁画《奉茶点》（局部）

　　仆人所端茶盘中大约是桃子。大定十四年（1174）墓葬，为金全盛时期墓葬。山西长子县小关村金代壁画墓。（长治市博物馆：《山西长子县小关村金代纪年壁画墓》，《文物》2008 年第 10 期）

第二章 宋、辽、夏、金、元时代茶文化及茶画制作
Chapter Two Tea Culture and Tea-Related Painting Creation of the Song, Liao, Xia, Kin and Yuan Dynasties

181

图 127

元·何澄（1223—?）·《归庄图》·吉林省博物馆藏品

　　作者皇庆元年（1312）被超擢为昭文馆大学士，此卷依据陶潜《归去来兮》描绘归来情形。元人绘画晋代人事，而所有屋舍器用皆为元代之物。备酒备茶，热烈而喜庆。卷后有赵孟頫等数位元代名家题款，弥足珍贵。

何澄·《归庄图》（局部）

图为三人合作点茶。

第二章 宋、辽、夏、金、元时代茶文化及茶画制作
Chapter Two Tea Culture and Tea-Related Painting Creation of the Song, Liao, Xia, Kin and Yuan Dynasties

183

图 128

元 · 刘贯道（1258—1336）**·《消夏图》·美国堪萨斯纳尔逊画廊藏品**

　　绢本设色，高 30.5 厘米，宽 71.7 厘米。

　　作者至元十六年（1279）绘《裕宗像》补御衣局使。有人以为画中原型人物为阮咸。人物头部有茶器、熏炉、阮弦，重屏里所绘人物正在侍弄高提手长流、带托茶盏等茶器，似乎为主人备茶。画中执壶形状与刘松年《斗茶图》中执壶非常相近。方形桌上有插花、卷轴、茶托、茶盏、三足鼎形盆景架。

图 129

元 · 刘贯道 ·《梦蝶图》

　　绢本，水墨设色。高 30 厘米，宽 65 厘米。

　　书案摆放与上图相似。

图 130

元·赵孟頫（1254—1322）·《斗茶图》

赵孟頫的《斗茶图》从人物的设计及其道具等的使用较多取自宋·刘松年的《斗茶图》，图中设四位人物，两位为一组，左右相对，每组中的有长髯者皆为斗茶营垒的主战者，各自身后的年轻人在构图上都远远小于长者，他们是"侍泡"或徒弟一类的人物，属于配角。图中左面这组，年轻者执壶注茶，身子前倾，两手臂向内，两肘部向外挑起，姿态健壮优美有活力。此图就茶道器而言，告诉我们这样的历史信息：1. 茶道形式为中晚唐流行起来的点茶道；2. 两组茶道小组都用三足风炉，虽然形制稍有不同；3. 点茶手是年轻伙计，年长的茶人的主要责任是品尝比赛对方的茶汤，观察其颜色变化；4. 烧水器不仅有他们身边的细长执壶，也有平底大肚带执手的茶壶——画面最右边炉子上的烧水壶；5. 画面左前方这个茶道师是整幅画的核心人物，他左手上举直腹白瓷茶杯，右手提着执壶竹蓝，他刚刚品了一口对方的茶汤，抿嘴回味茶味，对方师徒二人正全神贯注地看着他，等待他的评论，右方这位茶师左手上抬，好像在说："味道如何？"一副非常自信的神态，而他旁边的小伙计就显得拘谨多了；6. 在还没有完全结束右边这组茶汤的品评时，左边这一组已经开始点茶了；7. 从长幼分工可以推断：斗茶的核心是茶汤的滋味，点茶的动作好像不在比试范围之列，看谁家茶饼做得好；8. 在画面里找不出烘焙炉，也看不到敲砸碾罗茶饼的器具，茶粉好像是事前就已经准备好的。

第二章 宋、辽、夏、金、元时代茶文化及茶画制作
Chapter Two Tea Culture and Tea-Related Painting Creation of the Song, Liao, Xia, Kin and Yuan Dynasties

185

图 131

胡瓌·《卓歇图卷》·北京故宫
博物院藏品

　　人们静静地等待这位蒙古贵人对
这杯茶的审评！

图 132

元·倪瓒（1301—1374）·《雨后空林图》·台北故宫博物院藏品

　　画高 63.5 厘米，宽 37.6 厘米，1368 年作。

　　"空山新雨后，天气晚来秋。"雨后层林，水木清新，一杯绿茶，香沁心扉。空林寂静，更

倪瓒·《雨后空林图》（局部）

衬托出炉水的鼓浪腾波。

倪瓒，无锡人，初名珽，字元镇，号云林。是元代诗、书、画三绝归一的文化大家，与黄公望、吴镇、王蒙为元画坛四大家。宋建炎二年（1128），倪瓒先祖倪师道扈从宋高宗南渡居杭州，其长子倪子云于隆兴二年（1164）以吴县监丞的身份自杭州徙居吴县嘉定，次年"再徙无锡梅李乡之祇陀里，遂占籍"。后繁殖产业，为富一方而多慈善拯济，至父辈已经富足，产业更丰硕。倪瓒在养尊处优中寄情诗画，远离世俗，更对产业经营了无兴趣，被戏称为"倪迂""懒瓒"。加上梁溪倪门自宋泊元有隐逸遁世的传统，倪瓒在中年时代开始游历、漂泊的生涯，"舍北舍南来往少，自无人觅野夫家。鸠鸣桑上还催种，人语烟中始焙茶"。但人生遭际使他不得不面对俗务。1328年同父异母的兄长倪璨、1332年父亲文炳、1334年同父异母兄倪瑛相继去世，同辈中只余倪瓒一人协助伯父倪焕支应门户和庞大的产业。但是1345年，伯父倪焕去世，里里外外，独剩下倪瓒支撑家业。而长子倪诜的去世（1343—1347）对倪瓒来说是致命的打击。伯父去世令他深感独木难支，决计变卖家产，不留一缗，而孤舟漂泊。破败的家业由次子季民守护。1363年（癸卯）发妻去世，他更是万念俱灭，加上季民及其妻子不孝，倪瓒晚年极为苦寂凄婉。

画中有倪瓒题款："雨后空林生白烟，山中处处有流泉。因寻陆羽幽栖去。独听钟声思罔然。戊申三月五日云林生写。"从"寻陆羽幽栖去"等中老年的情绪看，此画当作于发妻去世五年后的1368年。作者大多数作品简洁而不设色，这是一幅少见的设色作品。作者在漂泊、幽居中逐渐从家道中衰、亲人相继离世的人生苦难中得到平复，归于平淡。在题款中汝易（陽）袁华有"门外青林生紫烟，龙泓一道落飞泉，恰如灵石山中宿。为说倪迂侣米颠"，将倪瓒与米芾相比。其另一幅未设色《安处斋图》有自题款："湖上斋居处士家，淡烟疏柳望中赊。安是为善年年乐，处顺谋身处处佳。竹叶夜香缸面酒，菊苗春点磨头茶。幽栖不作红尘客，遮莫寒江捲浪花。十月望日写《安处斋图》并赋长句，倪瓒。"这也是对倪瓒漂泊五湖三洲生活的写照。我们在图中找不到茶事的外在元素，但茶的宁静幽远却已掩压着心怀！幽栖人与茶烟相伴！

第二章 宋、辽、夏、金、元时代茶文化及茶画制作
Chapter Two Tea Culture and Tea-Related Painting Creation of the Song, Liao, Xia, Kin and Yuan Dynasties

189

图 133

元·佚名·《扁舟傲睨图轴》·辽宁省博物馆藏品

    画高 166 厘米，宽 111.9 厘米。

    在舟中读书品茗，何等快事！

图 134

元·朱德润（1294—

1365）·《林下鸣

琴图》·北京故宫

博物院藏品

琴音、茶香在

枯树水岸相益相尚。

图 135

元·朱德润·《秀野轩图卷》·北京故宫博物院藏品

　　在辽远清廓的天地间，远山近树衬托出一分安详，高阁之中，三位文友正在煮茶论道。

图 136

元·赵原·（？—1372）《陆羽烹茶图》（局部）

　　赵原，山东人，寓居苏州，擅山水画。图绘远山起伏，山水清远，水面辽阔，临溪筑有草阁，丛树掩映。阁内一人坐于榻上，当为陆羽，一童子拥炉烹茶。画中窥斑题款："睡起山垒渴思长，呼童剪茗涤枯肠。软尘落碾龙团绿，活水翻铛蟹眼黄。耳底雷鸣轻着韵，鼻端风过细闻香。一瓯洗得双瞳豁，饱玩苕溪云水乡。　窥斑。"作者自题款："山中茅屋是谁家，兀坐闲吟到日斜。俗客不来山鸟散，呼童汲水煮新茶。　赵丹林"，前有乾隆御题。

第二章　宋、辽、夏、金、元时代茶文化及茶画制作
Chapter Two Tea Culture and Tea-Related Painting Creation of the Song, Liao, Xia, Kin and Yuan Dynasties

193

图137

元·王蒙（1308—1385）·《品茶图》

王蒙为元末画家，赵孟頫外孙，湖州（今浙江吴兴）人。与黄公望、吴镇、倪瓒合称"元四家"。

峰顶古树苍茫，山腰庙宇隐于青松翠柏之间，山脚泉水潺潺，泉边茅草木亭内三人品茗，得茶之至味。三只白色茶盏置于桌上，三人结跏趺坐于草榻之上。

图中"公谅"题款："雾色如银莹白纱，梅葩影里月痕斜。家童乞火焚枯叶，漫汲流泉煮嫩茶。顿使山人清逸思，俄惊蜡炬发新花，幽情不减卢仝兴，两腋风生渴思赊。"

《元史》所载宇文公谅（1338年前后在世），字子贞，其先成都人。后徙吴兴。又有元代赵公谅，字允升，陕西栎阳人，元英宗至治三年（1323）为乡试举人，泰定帝元年（1324）中进士。是否是这两个公谅，待考。

邹中题款："嫩叶雨前摘，山斋和月烹。泉声云外响，蟹眼鼎中生。已得卢仝兴，复饶陆羽情。幽香逐兰畹，清气霭轩楹。"

蜀人黄岳题款："清泉细细流山肋（间），新茗丛丛绿芸色。良宵泊涧煮砂铛，不觉梅稍月痕直。喜看老鹤修雪翎，漫热沉檀捡道经。步虚声微茶初熟，两袖清风散杏冥。蜀人黄岳题于岷江寓所。"步虚为一种词牌，道家多用于其斋醮仪式。

杨慎题款："扁舟阳羡归，摘得雨前肥。漫汲画泉水，松枝火（皆）用微。香从几上绕，烟雨树头围。浑似松涛激，疑还绿绮挥。蜂鸣声仿佛，涧水响依稀。""绿绮"指代古琴，李白《蜀僧浚弹琴》："蜀僧抱绿绮，西下峨嵋峰。为我一挥手，如听万壑松。"自陆羽提出茶汤火候以"三沸"判定以来，茶人对壶中之水在不同温度下不同的声响有不同表述：陆羽以鱼目、蟹眼、腾波鼓浪来形容；苏轼以"松风"来形容。

图 138

元·盛懋（1341—
1370）·《秋 舸
清啸图》·上海
博物馆藏

　　盛懋主要活动
于至正年间(1341—
1370），湖光山水
间泛舟、品茶、读
经、吟诗，文人一
大雅趣也。

第二章　宋、辽、夏、金、元时代茶文化及茶画制作
Chapter Two Tea Culture and Tea−Related Painting Creation of the Song, Liao, Xia, Kin and Yuan Dynasties

195

图 139

元·佚名·《楼阁图》·辽宁省博物馆藏品

佚名·《楼阁图》（局部）

客来敬茶为南北风尚，品茗论道乃文
士雅趣。

第二章 宋、辽、夏、金、元时代茶文化及茶画制作
Chapter Two Tea Culture and Tea-Related Painting Creation of the Song, Liao, Xia, Kin and Yuan Dynasties

197

图 140

元·佚名·《竹林宴聚图轴》

　　在空旷的安静的空间里，三人品茶论道，古今中外无不在茶人的关照之下。

佚名·《竹林宴聚图轴》（局部）

第二章 宋、辽、夏、金、元时代茶文化及茶画制作
Chapter Two Tea Culture and Tea-Related Painting Creation of the Song, Liao, Xia, Kin and Yuan Dynasties

199

图 141

元·佚名·《百尺梧桐轩图》（局部）·上海博物馆藏品

原图高 29.5 厘米，宽 59.7 厘米。

主人回望茶水间，侍童已点好茶，正准备端过来！

图 142

元墓壁画《奉茶图》

元至元二年（1265），墓葬壁画，高约 118 厘米，宽约 152 厘米。

1958 年山西省大同市元代冯道真墓出土。已残毁。

墓向 188°。位于墓室东壁南侧。左边一虎眼石及牡丹花，右边几株毛竹，竹前为陈列茶具的方桌，贴有"茶末"标签的盖罐。倒置的茶盏、茶筅、叠摞的盏托、托盘、桃果等。中间站立一道童，左手握物于胸前，右手端一茶盏并托盏向前递出。这个壁画的一个重要价值在于告诉我们：茶末储存器的样式；大型茶事过程中，茶末是事前已经准备好的。至于现成的茶末是否成为商品尚有待考证。

第二章 宋、辽、夏、金、元时代茶文化及茶画制作
Chapter Two Tea Culture and Tea−Related Painting Creation of the Song, Liao, Xia, Kin and Yuan Dynasties

201

图 143

元墓壁画《侍女备酒图》·现存于长治市博物馆

    元至元十三年（1276），墓葬壁画，高约 159 厘米，宽约 176 厘米。

    2004 年山西省屯留县康庄村 2 号元墓出土。

    墓向 180°。位于墓室西壁。壁画有黑色边框，左侧有一高桌，桌上放置盖罐、碗、杯、劝盘、茶盏、汤勺、带荷叶盖瓷罐有可能是贮茶罐等。两名侍女站立于桌前，皆头梳红色花髻，身着红、黄色罗裙，一人持壶，另一人手捧劝盘。画面左上部有墨书题记"此位韩汝翼居中"。参考上图，荷叶盖罐应是贮存茶末之器物。

图 144

元墓壁画《侍女备酒图》·现存于长治市博物馆

元至元十三年（1276），墓葬壁画，高约 159，宽约 146 厘米。

2004 年山西省屯留县康庄村 2 号元墓出土。

墓向180°。位于墓室东壁上。壁画有黑色边框，右侧有一高桌，桌上放置茶罐、钵、茶盏托等。两名侍女立于桌后，皆头梳包髻，身着黄、红色罗裙，一持茶执壶，一持茶盏，茶筅击拂，后有一盘石茶磨。画面右上方有墨书题记，字迹漫漶不清，略可分辨为"此位堂韩赟五子至元"。点茶一般是两人进行：一人执壶点水，一人一手端盏，一手用茶筅击拂。

第二章 宋、辽、夏、金、元时代茶文化及茶画制作
Chapter Two Tea Culture and Tea-Related Painting Creation of the Song, Liao, Xia, Kin and Yuan Dynasties

203

图 145

元墓壁画《备茶图》

元大德二年（1298），墓葬壁画，高约 104 厘米，宽约 290 厘米。

1986 年山西省大同市齿轮厂 1 号元墓出土。已残毁。

墓向 179°。位于墓室东壁。中部一长桌，摆放茶盏托、注长流茶执壶、盖罐等；桌下一斗形盆装满瓜果等，一茶罐，其右侧有一株花草。左侧一女伎背身，抱一二弦琴；桌旁站立两名持物的侍女，右侧一女童端一盏托，旁边置一盆架，上置一盆。从花草生长情况看，茶事活动在室外进行。

图 146

元墓壁画《进酒图》（局部）

元至元二十五年（1288），墓葬壁画。

2001 年陕西省西安市韩森寨元代韩氏墓出土。现存于西安市文物保护考古所。

墓向 183° 。位于墓室西壁。为进酒图局部。该侍女为南起第三人，是该幅画面的中心人物。侍女面庞丰润，凤眼蛾眉、樱桃小口。包髻簪花、上着白色襦衫，下穿曳地长裙，肩披长帛，脚穿云头履。双手捧一红色大果盘，盘中置桃子、葡萄等水果。

第二章 宋、辽、夏、金、元时代茶文化及茶画制作
Chapter Two Tea Culture and Tea-Related Painting Creation of the Song, Liao, Xia, Kin and Yuan Dynasties

205

图 147

元墓壁画《侍女图》· 郑州市文物考古研究院藏品

墓葬壁画，高 102 厘米，宽 85 厘米。

1993 年河南省登封市王上村元墓出土。

墓向 190°。位于墓室东南壁，绘三侍女。左侧侍女身形高大，头梳圆髻，外着深蓝色褙子，足着尖头鞋，手捧一白盘，内盛两茶托盏。身旁一女，梳圆髻，戴耳环，着右衽窄袖襦，外罩半臂，下束曳地长裙，足着尖头鞋，执一团扇。此女身后一女，梳圆髻，左衽窄袖襦，外罩半臂，下束曳地长裙，足着尖头鞋，双手抱一葵花镜。执团扇妇女的出现，使人们自然想到，这是贵族家庭的生活。身后一童，左手持一玩偶，躲在大人身后。

图 148

元墓壁画《奉酒图》（摹本）·郑州市文物考古研究院藏品

墓葬壁画，高 102 厘米，宽 85 厘米。

1993 年河南省登封市王上村元墓出土。

墓向 190°。位于墓室西南壁，画右一女侍身形高大，头梳圆髻，外着蓝色褙子，足着尖头鞋，手捧白色劝盘，中放一小酒杯。中间侍女，梳双垂髻，着淡黄色左衽窄袖襦，外罩浅绿色半臂，下着百褶裙，足着尖头鞋，手捧黄色盘子，内盛杏、梨、瓜三种水果。左一女，梳圆髻，外着左衽窄袖襦，外罩白色半臂，下着曳地白裙，足着尖头鞋，双手抱一白色玉壶春瓶。

左图（图 147）奉茶，右图（图 148）奉酒。考古工作者的定名是正确的。随着茶事画的增多，人们对古代茶事了解不断深入。我们相信，对考古资料研究的深入，会令元代茶道文化以更加丰富多彩的面目呈现在世人面前。

第二章 宋、辽、夏、金、元时代茶文化及茶画制作
Chapter Two Tea Culture and Tea-Related Painting Creation of the Song, Liao, Xia, Kin and Yuan Dynasties

207

图 149

陕西蒲城洞耳村元墓壁画《献酒图》

其家具布置可资研究参考。

图 150

陕西蒲城洞耳村元墓壁画《歌舞图》

下图右后面双手端杯的人两个杯子大小形制不一，是否是酒之外还有奶、有茶？（资料出自刘恒武：《陕西蒲城洞耳村元墓壁画》，《收藏家》1999年第2期）

图 151

济南市历城区宋元墓壁画《夫妻对坐》

图 152

济南市历城区宋元墓壁画·《男女主人像》线图

　　桌上有食物、插花，桌下有酒罐，身后侍女端茶、酒。

第二章 宋、辽、夏、金、元时代茶文化及茶画制作
Chapter Two Tea Culture and Tea-Related Painting Creation of the Song, Liao, Xia, Kin and Yuan Dynasties

209

图 153

济南市历城区宋元墓壁画·《男女主人像》

　　埠东村石雕壁画墓北壁开芳宴壁画中的男女主人。（刘善沂、王惠明：《济南市历城区宋元壁画墓》，《文物》2005 年第 11 期）

图 154

M2 西北壁画宴饮摹本

　　男女主人公分用茶酒，左右两个杯盏形状大小不同，床前有
两酒瓶。图一有茶盏、茶托，图三有执壶。

第二章　宋、辽、夏、金、元时代茶文化及茶画制作
Chapter Two Tea Culture and Tea-Related Painting Creation of the Song, Liao, Xia, Kin and Yuan Dynasties

211

图 155

河北涿州元代墓壁画东壁《侍奉图》

　　右侧人从汤肴盆中舀茶汤，图中汤肴盆为内白外黑，是宋代茶事的普遍现象，很可能是一次性在汤肴盆中点好数人同时品饮的茶汤。（《河北涿州元代壁画墓》，《文物》2004 年第 3 期）

图 156

河北涿州元代墓壁画西壁《备宴图》

最前一人端盘，盘内两套茶托、茶盏，众人祭祀。

图 157

河北涿州元代墓壁画东南壁《孝义故事及云鹤图》

壁画中奉茶奉酒，奉食物。

第二章 宋、辽、夏、金、元时代茶文化及茶画制作
Chapter Two Tea Culture and Tea-Related Painting Creation of the Song, Liao, Xia, Kin and Yuan Dynasties

213

明代以『榷茶易马』为国策，对陕西、四川等地『边茶』出口采取严格管制，以提高国力。明代茶的政治地位提高，茶肆、茶坊比前朝有极大发展，新茶种层出不穷，茶品加工和烹瀹模式更新，茶学研究成果丰硕，茶事画的创作空前繁荣，佳作迭出。但是，明代茶道文化却没有唐宋时期厚重大气，帝王将相以及绝大多数茶学研究者，还没有把饮茶之道上升至治国修身的高度。以江南文士为核心的茶人集团是明代茶道主体力量，以书画界『明四家』为代表的士大夫不自觉地，但充分地体现了刚健率真、高雅清廉的茶人风骨。

第三章

明代茶道文化整体
评价与茶画创作

Chapter Three
Overall Evaluation of Tea Culture
and Tea-Related Painting
Creation of the Ming Dynasty

一、明代社会的整体风貌特征

与蓬勃进取的秦汉隋唐比较而言，明代是中国封建社会的稳健谨守阶段。同时，因其面临的特殊内外环境，明代也是中国社会急剧变化的时期。政治上，在吸收两宋败亡历史教训基础上，明代一方面加强皇权专制统治，另一方面又顺应中外交往日益频繁的趋势，采取开放的对外政策。在各方面，明代表现出与前朝的巨大差异。

首先，在政制体制上，大大压缩地方政府的数量，简化各级管理机构。明洪武初先后设置浙江、江西、福建、北平、四川、山东、广西、广东、河南、陕西、湖广、山西十二个行中书省。洪武九年（1376）改行省为承宣布政司（简称布政司），十五年（1382）又增加云南布政司。至此，全国共十三个布政司。布政司主行政，提刑按察使司（按察使）执掌一省之刑名按劾，都指挥使司（都司），掌卫所军政，合称都、布、按"三司"。"三司"互不统属，听命中央的直属上级。行省下设府、州、县三级行政区划。

A. General Feature of Tea Ceremony Culture of Ming Dynasty

The Ming Dynasty was a steady and self-controlled stage in Chinese feudal society compared with vigorous and enterprising Qin, Han, Sui and Tang dynasties at the same time was a period of facing rapid social changes due to its special internal and external environment. The central government of the Ming Dynasty on the one hand strengthened authoritarian imperial power on the basis of absorbing history lesson of the destruction of the Northern and Southern Song dynasties, on the other hand adopted open-to-the-outside-world foreign policy complying with the trend of the growing communication between China and foreign countries. So the Ming Dynasty differs enormously from the former dynasties in all aspects.

Firstly, the huge differences between the Ming Dynasty and the former dynasties were as following: decreased on a large scale the number of local government and simplified the management institutions in terms of political system. Twelve provinces were successively set up including Zhejiang, Jiangxi, Fujian, Beiping, Sichuan, Shandong, Guangxi, Guangdong, Henan, Shaanxi, Hubei, and Shanxi during the early period of the Hongwu Reign of the Ming Dynasty. "Chief Secretary" was began to be used in stead of "Province" in 9th year of the Hongwu Reign (1376). Yunnan was established in 15th year of the Hongwu Reign, thus there were 13 Chief Secretaries (Buzhengsi) at that time in China. The Chief Secretary primarily responsible for administration, the Provincial Judge Office (Anchashi) mainly responsible for criminal law and criminal cases, the Chief Provincial Military Leading Organization (Dusi) responsible for military affairs, these

其次，在社会生活内容上呈现出多元多层次状态。中古时期的士、农、工、商"四民"在明代成为"六民"，增加了兵、僧。到了明代中后期，姚旅才提出"二十四民"，在"六民"之外，增加"道士、医者、卜者、星命、相面、相地、弈师、驵侩（骡马交易经纪人）、骂长、舁夫、篦头、修脚、修养、倡家、小唱、优人、杂剧、响马贼"等"十八民"。人们普遍看重功名利禄，追求自我物质享受。

社会生活空间的变化较快。一是明政府的迁都三京：吴元年（1367）徙苏州富民实濠梁。明初，又迁苏、松、杭、嘉、湖民之无田者，往耕临濠，官给牛、种，免赋三年，明初徙江南富民 1.4 万户到中都。定鼎南京后，徙四方巨族实之，太祖命户部籍浙江等九省及应天 18 府富民 14300 户，依次召见，并迁至南京。成祖即位，徙太原、平阳、泽潞、辽、沁等府丁多而田少及无田者，以实北京，迁直隶浙江之民 2 万户，充仓脚夫；徙浙江、应天府富民 3000 户充当北京宛平、大兴二县之厢长，附籍北京，仍应本籍徭役。二是社

three departments did not have subordinative relationship but just took orders from their direct superiors of the central government. Fu (Government Office), Zhou (Prefecture) and Xian (County) were three inferior administrative divisions under the provincial level.

Secondly, the content of social life was in a multivariate and multilevel status. Soldiers and Buddhist monks became two new social classes besides of other four types including scholars, farmers, craftsmen and merchants which had been the four major social classes from the medieval times. In the middle and late periods of the Ming Dynasty, Yao Lucai put forward his view of twenty-four social classes by adding Taoist priests, doctors, diviners, fortune-teller by astrology, fortune-teller by physiognomy, fortune-teller by landscape, chess players, horses trading agents, bearers, hair combers, pedicures, prostitutes, ditty singers, clowns, poetic drama performers, robbers, etc. apart from the above-mentioned six social classes. People valued fame and fortune and pursued for self material comforts.

The social life space changed quickly, and it was firstly showed by three-times moving of the capital of the Ming Dynasty and related immigrations. The fiest time, rich families of Suzhou were moved to Haoliang in 1368, landless people from Suzhou, Songjiang, Hangzhou, Jiaxing, and Huzhou to Linhao in the early Ming Dynasty. Cattle, seeds, and three-year-tax were free to immigrants. 14000 rich families from south-of-Yangtze-River regions were moved to the capital. The second time, rich and powerful 14300 families from 9 provinces including Zhejiang and other 18 prefectures including Yingtian were moved to Capital (Nanjing). The last time, landless people or few-land owners from Taiyuan, Pingyang, Zelu, Liao, Qin prefectures, and 20000 families from Zhili and Zhejiang provinces were moved to Beijing of whom the latter were made to work as porters in the Chengzu

第三章 明代茶道文化整体评价与茶画创作
Chapter Three Overall Evaluation of Tea Culture and Tea-Related Painting Creation of the Ming Dynasty

217

会流动加快，军屯人口增加。朱元璋起事时，就注重军屯。军士随军征讨，军屯不断变化，战事减少直至完全统一后，军士及其家属随军屯驻留，形成新的农民。士人、商人、僧道人员、工匠艺人等的迁流也很频繁。此外，明初几次强力降低人身依附关系，也加速了人员的自由流动。

最后，人的自我意识普遍彰显。处在流动变化中的个人、家庭、家族或小团体，往往采取步步为营、量入为出的稳妥自保原则处事、置产，势利世故是必然的普遍现象。庠序之中，老师为了束脩而一味顺从庠生，师道尊严名存实亡。唐李白"千金散尽还复来"的自信，梁山豪杰四海之内皆兄弟的胸怀，在物质利益面前已经成为梦中呓语。主衰遭奴欺成为普遍现象，得意门生不念昔日座主亦非个别案例。书院、寺宇、道观也不宁静。道士方技的炼丹术比他们的修养重要，他们为富贾大族歌功颂德比探讨学问重要。地方惠利高于中央统一要求。加上明代帝王多无能喜欢奢靡且年寿不高，助长了明代社

Reign; 3000 families from Zhejiang and Yingtian Prefecture were moved to Wanping and Daxing counties of Beijing but these immigrants still had to take corvee in name of their original living places. The quick change of social life space showed secondly by the accelerated social mobility is that military-tillage-concerned population increased, these people turned into farmers after the war completely ceased; soldiers, merchants, Buddhist monks, Taoist priests, craftsmen and artists also moved frequently. Furthermore, early-Ming's several policies of reducing personal dependence played a role in accelerating the free flow of people.

Thirdly, the huge differences between the Ming Dynasty and the former dynasties were reflected by the increase and expressing of self-awareness of common people. Individuals, families, clans or small groups often handled issues such as dealing with others and buying estates by taking step-by-step and pay-as-you-go as self-preservation principle in the flowing and changing circumstance, so snobbish was the inevitable phenomenon. In school, teachers were blindly obedient to students just for the sake of payment for private tutor, the teacher's dignity existed in name only. Li Bai's Tang-Dynasty-type confidence like totally-dispersed-money will surely return and Northern-Song-Dynasty-type chivalrous heart of Liangshan heroes' treating all friends within the whole country as brothers have become dream words in front of the material benefits. Servants insulted their masters when the masters' family fortunes declined and favorite pupil lostgrateful heart towards their teachers had not been exceptional cases in the Ming Dynasty. The academy of classic learning, Buddhist temples, Taoist temples were not pure land anymore, Taoist priests regarded medicines and spagirism more important than their self-cultivation, praising rich and powerful families was more important than studying knowledge. Local profit was

会的自私、扬厉、狂躁、贪婪、奢靡风气的弥漫。纵观有明一代，位列政要而清谨廉洁如于谦者凤毛麟角。但在这些急剧变化面前，一批文人士大夫，谨守儒家理念，矜持诗人节操。饮茶以前代所没有过的频率和密度出现在画坛大师的丹青之中，茶事成为画坛流行题材。以画坛"明四家"作品为代表的明代茶事画近60幅，占唐宋以来中国茶事画的一半以上。

经济生活中，社会分化比之前代更为突出。皇亲国戚和望族富户，以不足百分之五的人口，占有全国百分之七十以上的土地。这些人连同大商人，控制着国家经济。以丝织业和棉纺织业为产业链核心的工商业发展迅速，形成苏、湖、松江等中心。这些地方也成为宋末以来城镇化的中心。新兴城市集镇多在江南茶乡出现。茶肆茶坊遍布这些城镇之中，饮茶风尚极为普遍。

在对外关系方面，明朝面临的主要压力来自蒙古和南方的海上倭寇。明朝对北方马匹的需要，对海外香料等奢侈品的需要，以及蒙古、日本等国家或政权对中

more significant than the unified requirements from central government and short-life emperors of the Ming Dynasty were mostly incompetent but like luxurious life having contributed to the social customs of selfish, manic, greed and extravagant. The number of politician like Yu Qian who was in an important post in government but executed their duty with cautious, honest and upright was too small throughout the Ming Dynasty. A batch of literati strictly adhered to the Confucian philosophy and reserved poets to the moral integrity in front of these dramatic social changes. Tea-drinking images appeared in an unprecedent frequency and density in paintings of master painters thus tea activity inevitably became the popular theme in painting circle. Nearly 60 pieces of tea-activity paintings represented by those created by four master painters of the Ming Dynasty, accounted for more than half of Chinese tea-activity paintings from the Tang and Song dynasties.

More prominent social differentiation than the previous generation occurred in the economic life. Princes, princesses, and relatives of royal family, distinguished and rich families with less than 5% of the total population owned more than 70% of the country's land. These people together with powerful businessmen controlled the national economy. Silk-and-cotton-textile industry and commerce centered in Suzhou, Huzhou, Songjiang etc., as the core of the industrial chain developed rapidly. These places had also become the center of urbanization since the Song Dynasty. The emerging cities and market towns appeared in the major tea producing areas. Tea markets and tea houses located all over the cities and market towns made tea drinking an extremely common fashion.

The main pressure the Ming Dynasty government faced came from the Mongolians and Japanese sea pirates in southern

第三章 明代茶道文化整体评价与茶画创作
Chapter Three Overall Evaluation of Tea Culture and Tea-Related Painting Creation of the Ming Dynasty

219

国丝绸、瓷器、茶叶、铁器及食品等的需要，已经持续化、常态化。万历《大明会典》记载的各国进贡品达四十余种。到万历十七年（1589）《陆饷货物抽税则例》列举的货物超过百余种。近年以来，以"南海一号"打捞与揭取为标志的中国水下考古事业的开展，使中国宋元明清时代的对外贸易向人们呈现了一个更为壮丽的古代图景。在数以千计的完整的出水明代白瓷及青花瓷中，我们看到精美的执壶、茶盏等茶道器。自唐代"安史之乱"以来，中国中央政府一直采用专卖形式用中国茶叶换取草原民族的马匹。先是同回鹘，后来同吐蕃、辽、夏进行茶马贸易。明朝主要同蒙古（鞑靼）、乌斯藏（吐蕃）、维吾尔进行茶马贸易，实行"榷茶易马"政策。另外，中国政府将茶叶也作为一种战略资源用之于对西藏和西域的外交控制。

二、"榷茶易马"成为明代一项国策

从各种信息来分析，明代作为一个统一的多民族中央集权制

China in terms of foreign relations. It had been normalized that the Ming Dynasty government needed to import horses from the north and spices from overseas also export Chinese silk, porcelain, tea, iron ware, and food to those related countries. According to the records in "Code of Great Ming Dynasty" compiled in the Wanli Reign of the Ming Dynasty, the tributes from other countries to Ming Dynasty court were more than 40 kinds, over 100 in 17th year of the Wanli Reign (1589) based on the statistics from *Cases of Reimbursement of Levying Tax on Land-transported Goods*. In recent years, the start of China underwater archaeological career marked by re-floatation of "No.1 South China Sea" sunken ship unfolded us a more magnificent picture of China's foreign trade in the Song, Yuan, Ming, and Qing dynasties. Tea wares including ewers and calyxes were found among thousands of well-preserved white porcelain, blue and white porcelain of the Ming Dynasty in "No.1 South China Sea" sunken ship. Since the "An-Shi Rebellion" of the Tang Dynasty, China's central government has used monopoly policy of sale to get horses from horse-riding races including the Huihus (ancestor of Uygur), the Tubos (ancestor of Tibetan), Liao, and Western Xia regimes with Chinese tea. The central government of the Ming Dynasty mainly had been engaged in tea-horse trading with the Mongolians, the Tubos, and the Uygurs. In addition, the Chinese government took tea as a kind of strategic resources to control Tibet and the Western Region.

B. "Taxation and Monopoly of Tea and Tea-horse Interchange Trade" Turning into National Policy of Ming Dynasty

The Ming Dynasty, as a unified, multi-ethnic and centralized country, had created conditions for domestic

国家，无疑为国内外的交通、交流创造了条件。茶叶生产和消费继续呈现向前发展的态势。14世纪60年代后期，蒙古军队被赶出中原后，明军对马匹的需求变得突出而迫切，最初在西南的四川、云南和贵州，通过茶、盐和丝织品的专卖来换取马匹，加以解决。每个茶马市场，100万斤茶换取1.4万匹马。到了1387年，山西行省成为主要茶马交易市场。1421年，首都迁至北京后，中国传统的以北方草原马匹为主要对象的政府采购加快了。1407—1427年，从朝鲜购入了1.8万匹马[1]。但是，英宗正统十四年（1449）被俘虏的"土木堡事件"使边境贸易骤然停顿。宪宗成化六年（1470）与蒙古的边境贸易得以恢复。

1. 明初延续前代"榷茶制度"以马易茶

如果大家再重新阅读一下《明太祖实录》，一定会发现，明太祖在位后期（洪武二十五年至三十一年），在儒家士大夫帮助下，大明政权的各种礼仪（祭祀、加冕、外番朝拜）完备以后，他除了对

and foreign transportation and exchange according to the analysis based on various information. Tea production and consumption developed continuously. In the late 1360s, the Ming Dynasty army's demand for horses suddenly turned into an urgent one after Mongolian army of the Yuan Dynasty was driven out of the Central Plains of China. So the central government of the Ming Dynasty tried to get horses in monopoly exchange for tea, salt and silk in southwest China's Sichuan, Yunnan and Guizhou. In every tea-horse market, any one could get 14000 horses in exchange for 1 million catties (1.5 million pounds) tea. Shanxi Province was the most important tea-horse trading market in 1387. The central government strengthened monopoly to purchase of horse from north China after the capital moved to Beijing in 1421. 18000 horses were bought from Korea in 1407-1427.[1] However, the border trade stopped abruptly due to the occurrence of the Tumubao Event in 14th year of the Zhengtong Reign (1449) of Yingzong Emperor of the Ming Dynasty. The border trade with Mongolia was resumed in 6th year of the Chenghua Reign (1470) of Xianzong Emperor of the Ming Dynasty.

1. "Taxation and Monopoly of Tea and Tea-horse Interchange Trade" Policy Was Continuously Executed in the Early Ming Dynasty

You will find such description if you have time to read *The memoir in Taizu Emperor's Reign of the Ming Dynasty* that after the completion of etiquette such as sacrificing, coronation, foreign diplomats' pilgrimage, etc. With the help from Confucian scholars in the late period (25th to 31st year of the Hongwu Reign) of the Hongwu Reign of Taizu Emperor of the Ming Dynasty. Taizu Emperor took some measures to reinforce his military deployment and military officials allocation, issued

第三章 明代茶道文化整体评价与茶画创作
Chapter Three Overall Evaluation of Tea Culture and Tea-Related Painting Creation of the Ming Dynasty

221

军事部署和武官调配特别重视以外，对关系国计民生和综合国力的茶马互市也特别关注，并有明确而具体的诏令。他历史性地废除了唐末宋初以来最昂贵的茶叶"龙团凤饼"，改为芽茶进御：

庚子诏建宁岁贡上供茶，听茶户采进，有司勿与。敕天下产茶之处，岁贡皆有定额，而建宁茶品为上，其进必碾而揉之压以银板大小龙团。上以重劳民力，罢造龙团，惟采茶芽以进。其品有四：曰探春屯春次春紫笋。置茶户五百，免其徭役，俾专事采植。既而有司恐其后时，常遣人督之，茶户畏其逼迫，往往纳赂，上闻之故有是命。[2]

次年"五月辛卯福建建宁府民贡品茶六百余斤"[3]。唐宋贡茶往往以穿（串）为计算单位，分为大串小串。到洪武二十五年五月开始进奉芽茶，也就是炒青散茶。对茶叶生产和"榷茶易马"，中央政府直接督查、管理：

洪武二十九年三月戊午朔，遣官往四川天全六番招讨司核实洪武二十四年至二十六年茶课，己丑长兴侯耿炳文奏秦州（治今甘肃天水市）茶马司不便于互市请迁于西宁，

clear and concrete imperial edict on tea-horse interchange trade which relates to people's livelihood and comprehensive national strength very much. Taizu Emperor also abolished the most expensive "Coiled-Dragon-and-Phoenix Pie" tea which had been made since the late Tang and early Song dynasties using bud-tea instead as the tribute to the royal court." In 24th Year of the Hongwu Reign (1391), Zhu Yuanzhang (Taizu Emperor) issued an imperial edict to abolished 'Coiled-Dragon-and-Phoenix Pie' tea which made in the former dynasties as the tribute to the royal court. Such annual tribute was prepared in a fixed amount by the tea producer without disturbing from local officials. Jianning tea was of the highest quality among all tea tributes which was made by grinding and massaging then pressing with silver plate into a round 'piece as coiled-dragon-and-phoenix pie', and this process required a huge amount of labor force. So no more 'Coiled-Dragon-and-Phoenix Pie' tea should be made and only bud tea of four grades including Tanchun, Tunchun, Cichun, and Zisun can be the annual tea tribute to the royal court. Five hundred families will be selected as royal tea tribute producers who will be allowed to avoid providing the corvee."[2]

In the following year, "more than six hundred catties Jianning tea from Fujian Province as the annual tea tribute to the royal court was received on May 28th".[3] Chuan (namely string) falling into Big Chuan and Small Chuan had been the unit for measuring the royal tea tribute during the Tang and Song dynasties. The 25th Year of the Hongwu Reign (1392) was the first year for the royal tea tribute producer to offer tea leaves instead of tea pie to the royal court. The central government handled the direct supervision and management over tea production and tea-horse interchange trade. In March 1396 (29th Year of the Hongwu Reign), officials were sent

命户部议之。[4] 洪武三十年三月癸亥，遣驸马都尉谢达往谕蜀王椿曰：秦蜀之茶自碉门黎雅至朵甘乌斯藏五千余里皆用之。其地之人不可一日无此。迩因边吏稽查不严，以致私茶出境为夷人所贱。夫物有至薄而用之，则重，茶是也。始于唐而盛于宋。至宋而其利博矣。前代非以此专利，盖制夷狄之道当贱其所有而贵其所无耳。我国家榷茶本资易马，以备国用。今惟易红缨杂物，使番夷坐收其利，而马入中国者少，岂所以制夷狄哉？尔其谕布政使都司严为防禁无致失利。[5]

对茶马互市的有关奏请，洪武皇帝都给予高度重视，并很快明确答复：

洪武三十年四月己丑改秦州茶马司为西宁茶马司迁其治于西宁，从长兴侯耿炳文之请也。[6]

2. 对于破坏"榷茶易马"国策的人员给予严厉惩处

驸马都尉欧阳伦成为明代历史上第一个因违反"榷茶易马"制度贩鬻私茶而被处极刑的皇亲国戚：

洪武三十年六月乙酉驸马都

to Tianquan in Sichuan in order to verify the tea tax from 1391 to 1393 (24th to 26th Year of the Hongwu Reign). Soon afterwards Geng Bingwen—Changxing Marquis sent a document to the emperor saying that Qinzhou Tea—horse Bureau should move to Xining for management convenience, Ministry of Revenue discussed it.[4] In March 1397 (30th Year of the Hongwu Reign), Xieda, a commandant—escort, was sent to instruct Chun who was King of Shu that tea produced in Qin and Shu including the five—thousand—Li area from Diaomen, Liya to Duogan and Wusizang will all be levied. The smuggling tea to other countries were sold at a lower price because the frontier officials did not strictly check the trading activity. Although tea leaf is so thin and light, it is important because it is very useful. The tea trade started from the Tang Dynasty, prospered and became profitable in the Song Dynasty. There was no "Taxation and Monopoly of Tea" policy in the former dynasties. It was the diplomatic strategy of the Ming Dynasty to make it important and raised its selling price to other countries by means of government policy, getting horses in need in monopoly exchange for tea was also a national strategy. But now what we get through trade from other countries are only cats and dogs not horses we need very much while other sides benefit from it well. How can we conquer other tribes? So provincial administrative government should discharge its duty strictly to prevent the loses in trading.[5]

Taizu Emperor paid a high attention to the document submitted by Geng Bingwen—Changxing Marquis and answered him very soon. "Qinzhou Tea—horse Bureau will be renamed as Xining Tea—horse Bureau on April 26th, 1397 (30th Year of the Hongwu Reign) and its office can be moved to Xining as Changxing Marquis' request for management convenience."[6]

第三章 明代茶道文化整体评价与茶画创作
Chapter Three Overall Evaluation of Tea Culture and Tea—Related Painting Creation of the Ming Dynasty

223

尉欧阳伦坐贩私茶事觉赐死。初，上命秦蜀岁收巴茶听西番商人以马易之，中国颇获其利。其后商旅多有私自贩鬻，至为夷人所贱，马价遂高。乃下令严禁之。有以巴茶出境者实以重法。伦尝遣家人往来陕西贩茶出境货鬻，依势横暴，所在不胜其扰。虽藩阃大臣皆畏威奉顺，略不敢违时。四月农方耕耨。伦适在陕西令布政使移文所属，起车载茶往河州（治今甘肃临夏回族自治州），伦家人有周保者尤纵暴，所至驱迫有司索车五十辆至兰县河桥巡检司，捶辱其吏，吏不能堪，以其事闻。上大怒，以布政使司官不言并伦赐死。保等皆坐诛，茶货没入于官。以河桥吏不避权贵，遣使赍敕嘉劳之。[7]

可能与锦衣卫、按察使的积极努力有关，洪武皇帝对榷茶易马的具体情况比较熟悉，往往会发出具体而符合实际情况的诏谕：

（洪武三十年七月）辛酉命户部于四川重庆保宁三府及播州宣慰使司置茶仓四贮茶以待。客商纳米中买及与西番商人易马，各设官以掌之……辛未上谓户部

2. Severe Punishment for Destruction Deeds towards National "Tea-horse Interchange Trade" Policy

Ouyang Lun, a commandant-escort, was the first relative of the emperor in the history of the Ming dynasty who was put to death for violating "Tea-horse Interchange Trade" policy by selling private tea.

Ouyang Lun, a commandant-escort, was forced to commit suicide for selling private tea on June 22nd, 1397 (30th Year of the Hongwu Reign). At the beginning, the central government gave order allowing Shaanxi and Sichuan to sell their local tea products to other countries and regions to get horses which was a good profit. Many merchants afterwards sold private tea so that the tea price decreased while the horse price increased, thus the central government ordered to ban private selling of tea. Those who dare to sell Sichuan Tea to other countries and regions will be seriously punished. Ouyang Lun has tried to send family member to Shaanxi buying tea in there perversely and violently and sell it abroad, even the local officials feared his power and offered co-operation timidly without any violation. Farmers began to plow and weed in April. Ouyang Lun was just in Shaanxi and urged the officials of Shaanxi Provincial Chief Secretary providing him vehicles to carry tea to Hezhou. One of Ouyang Lun's family members Zhou Bao forced officials cruelly on their way to extort 50 carts. Zhou Bao beat the officials from Social Security and Public Affair Department when they arrived at Heqiao in Lan County. Those officials could not bear it any more and reported the situation to superior departments. The superior leaders were very angry about such affair and sentenced Ouyang Lun and Zhou Bao to death due also to the fact that the officials of Shaanxi Provincial Chief Secretary had not reported it in time. At the same time

尚书郁新等曰：陕西汉中以茶易马每马约与茶百斤，给马三百万斤可易马三万匹，宜严守关隘禁人贩鬻。其四川松茂之茶与陕西同，碉门黎雅则听商人纳米市易。尔户部即遣人于陕西、四川按视茶园之数。[8]

次年四月，户部在成都、重庆、保山和播州宣慰使司设立了四个存贮茶叶的仓库。

对具体出现的外番违法问题，洪武皇帝也进行妥善处理：

（洪武三十年七月）丁酉兰州奏朵甘乌斯藏使臣以私茶出境，守关者执之请寘以法，上曰禁令以防关吏及贩鬻者，远人将以自用，一时冒禁勿论。[9]

虽然当时人们没有外交豁免权这个概念，但洪武皇帝从大局出发，对西藏使臣法外开恩。这有利于边贸发展和相互合作。在茶马互市中，我们看到贸易点向外延伸的同时，对茶叶生产管理越来越严格。陕西布政使司的地位越来越高。洪武皇帝对陕西也极为重视，秦王、太子先后驻跸西安。浙江是当时人口最多的地区，而陕西可能是面积最大的

dispatched envoys to grant rewards to the officials from Heqiao Social Security and Public Affair Department of Lan County and confiscated Ouyang Lun's tea.[7]

Taizu Emperor could often issue concrete and reasonable imperial edits on tea-horse interchange trade based on accurate and timely reports from Jinyiwei— Emperor's Guards Forces and Provincial Judge Offices:

In July 1397 (30th Year of the Hongwu Reign), Ministry of Revenue ordered Sichuan, Chongqing and Baoning government offices and Bozhou National Autonomous Prefecture built four tea warehouses in case that merchants would buy or be in exchange for getting horse, officials were assigned to take corresponding responsibility. Yu Xin, Minister of Ministry of Revenue once said that one can exchange one hundred catties tea for one horse, three million catties tea for thirty thousand horses in Hanzhong of Shaanxi, the border pass should be closely guarded to avoid tea selling. Tea produced in Songmao of Sichuan is the same as that from Shaanxi, merchants can do the deal in the market of Diaomen and Liya. The Ministry of Revenue sent officials to inspect tea gardens in Shaanxi and Sichuan so as to know the total number of tea garden.[8]

Four tea warehouses were built in Chengdu, Chongqing, Baoshan and Bozhou National Autonomous Prefecture by Ministry of Revenue in April 1398 (31st Year of the Hongwu Reign).

Taizu Emperor could often properly deal with related illegal problems: A document submitted in July 1397 (30th Year of the Hongwu Reign) from Lanzhou saying that envoys from Duogan and Wusizang left the country with their private tea, the border pass guarders approached and told them that border-pass officials and sellers were forbidden to take tea

第三章 明代茶道文化整体评价与茶画创作
Chapter Three Overall Evaluation of Tea Culture and Tea-Related Painting Creation of the Ming Dynasty

225

区域，而且处在面对蒙古和回族及撒马尔罕等的最前沿。陕南茶叶也受到明朝政府的重视。洪武十七年（1384），四川布政使所辖乌萨乌蒙芒部年易马 6500 匹；十九年九月，行人（商人）冀中一次从陕西易马 2807 匹。"十月己卯陕西庆阳等卫将士四百九十余人送马至京，赐钞有差。"[10]

近 3000 匹马，需要 500 人运送，人马吃喝践踏，其成本也相当高。

3. 国家统一，交通网络初步形成，为茶马互易创造条件

洪武二十七年（1394）九月官修《寰宇通衢书》完成，对全国的交通线路进行划分、登记。把原来 47 道压缩为 41 道，从南京出发计算行程，到云南 6000 多里，到陕西、北平都是 5000 多里。到陕西要经过 12 个布政使司。

从洪武皇帝晚年的言行看，在采纳有关建议后采取唐代京师以重驭轻的军队部署方针，陕西、山西、河南、北平等边疆地区的军队数量几乎减少一半。陕西军队从 23 万减少到 10 万。同时五军所在都进

abroad, then tried to check it according to legal rule, those envoys said they took the tea for their own use and it was regardless of the prohibition.[9]

Although people at that time had no concept of diplomatic immunity in the Ming Dynasty, Taizu Emperor endowed envoys of Tibet this privilege from the general interest which benefits the development of border trade and inter-state cooperation. Outward extending of the trade and more strict management on tea production were seen in the tea-horse interchange trade. Chief Secretary of Shaanxi Province shared a higher and higher position. Taizu Emperor took Shaanxi very seriously, the king of State of Qin, prince of Qin lodged successively in Xi'an. Zhejiang was the most populous region, and Shaanxi was the largest one in the face of Mongolia, Hui nationality areas, Samarkand, etc. Tea produced in southern Shaanxi was valued highly by central government of the Ming Dynasty. 6500 horses were trading in Wusa, Wumeng and Mangbu under the jurisdiction of Sichuan Chief Secretary in 1384 (17th Year of the Hongwu Reign). One merchant called Ji Zhong exchanged 2807 horses in Shaanxi at a time in September 1386. "On October 16th, 490 officers and soldiers from Shaanxi and Qingyang departed to escort horses in transportation to the capital—Beijing were given money there differently."[10]

The cost including food and drink for 500 officers and soldiers to escort nearly 3000 horses in transportation were quite high.

3. National Unification and Initially-Shaped Transportation Network Creating Conditions for Tea-horse Interchange Trade

The officially compiled The National Road Manual was

行军屯，其中20个卫所牧养军马，同样军马不进入市场交易，即使病弱之马也要养在军中。

和平时期，执行严格的"榷茶易马"成为明代一项国策。以贵我所有而贵人所无，贱我所无也是贱人所有。与之相关，洪武二十七年七月"甲寅禁民间用番香番货……两广所产香木听土人自用"[11]，而不能出境交易。对耗费大量白银的南洋香料进行严格控制，民间不得使用。

因为将"榷茶易马"作为一项国策，故以官方为主体，以经济贸易入手，以提高外交影响和增强国力为目标的茶叶外销，自然将茶叶提高到前所未有的地位——高过了丝绸与瓷器。

三. 茶叶政治文化地位空前提高

我国南方很早就有用茶祭奠的习俗。最早文字大约出现于《南齐书》卷三。《茶经·七之事》云："南齐世祖武帝遗诏：'我灵座上慎勿以牲为祭，但设饼果、茶饮、干饭、酒脯而已。'"但这只属于个例。中晚唐时期，《茶经》

finished in September 1394 (27th Year of the Hongwu Reign) in which we can find the division and registration of the traffic lines within the whole country. The original 47 traffic lines were compressed to 41, the distance from Nanjing to Yunnan was more than six thousand Li (1 Li = 0.5 km, more than five thousand Li from Nanjing to both Shaanxi passing through 12 Chief Secretaries and Beijing.

In his late years, Taizu Emperor adopted the proposal of taking Tang-Dynasty guidelines of making the capital heavy in order to control the light troops, so the number of troops stationed in Shaanxi, Shanxi, Henan, Beijing, and other border areas was almost cut in half. The Shaanxi troop was reduced from 230000 to 100000 people. The above-mentioned troops carried out military tillage in their stationed places and raised war-horse at 20 garrisons as well. But war-horse was not allowed to entered the market and even the sick horse would have to be raised in the army.

In times of peace, the strict implementation of "Taxation and Monopoly of Tea and Tea-horse Interchange Trade" had become a national policy in the Ming Dynasty. Cherish what we have, Despise what we don't have. Associated with this view, "Imported incense and goods were forbidden to used in the society... scented wood produced in Guangdong and Guangxi could only be consumed locally."[11] and not allowed to be sold abroad from July 1394 (27th Year of the Hongwu Reign). Spices from Southeast Asia were strictly controlled.

Because of the national policy "Taxation and Monopoly of Tea and Tea-horse Interchange Trade", governmental authority as the main body dealt with tea export aiming at improving the foreign influence and national strength through economy and trade, hence tea naturally shared an unprecedented high position which is higher than that of silk

第三章 明代茶道文化整体评价与茶画创作
Chapter Three Overall Evaluation of Tea Culture and Tea-Related Painting Creation of the Ming Dynasty

227

面世，茶道文化大行天下。李郢《茶山烘焙歌》言"十日王程路四千，到时须及清明宴"是对湖州长兴贡茶院为赶上清明宴而日夜加工供紫笋茶情形的生动描写，也揭示了唐宫廷茶宴的实际存在。法门寺地宫出土一套宫廷茶器。而在宋代的文献中，我们看到在接待辽、金、夏使节时，使用茶汤。元代，饮茶也是贵族生活所不可缺少的一部分。但在文化上的地位毕竟有限。入明以后，茶的地位陡然提高。这与朱明开国君臣生活在茶乡、政权初期以苏浙为腹地的现实密切相关。

元年三月丁丑朔，宣州供新茶，上命内夫人亲煮荐于宗庙。[12]

洪武二年五月辛酉诏，凡时物，太常先荐宗庙，然后进御。[13]

洪武二年六月丁亥，造太庙神器成。每庙壶一盂一台盏二爵二栀四樏十橐四匙二茶壶二钟二炉一香合一花瓶二烛台二，计金八千八百八十两。先是，上欲造太庙金器因谕礼官曰："礼缘人情不必拘泥于古。近世祭祀皆用古笾豆之属。宋太祖曰：'吾先祖亦不识此。'孔子曰：'事死

and porcelain.

C. Unprecedented High Political and Cultural Status of Tea

The custom of offering tea as sacrifice has a very long history in southern China. The earliest related written record might be that in volume 3 of *Book of Southern Qi Dynasty*. One example can be found in "Tea Activity" from *The Classic of Tea*, Shizu Emperor of Southern Qi Dynasty wrote in his testamentary edict. "Do not put animal sacrifice on the pedestal of my coffin but cakes, tea, dried food, alcohol, and preserved fruits." *The Classic of Tea* emerged in the middle and late Tang Dynasty, tea culture thus popularized nationwide. Li Ying wrote in *Tea Baking Song*, "Four thousand Li should be covered in 10 days in order to offer tea for the Tomb-sweeping Day royal feast" describing vividly day-and-night processing of purple bamboo tea at Changxing Tribute Tea Institution in Huzhou also revealing the existence of palace tea feast in the Tang Dynasty. A set of royal court tea ware was unearthed from the underground palace at Famen Temple. We have found record in the Song-Dynasty document on receiving envoys of Liao, Kin, and Western Xia regimes with tea as drink. Drinking tea was an indispensable part of the aristocratic life in the Yuan Dynasty, but its cultural status after all was limited. The sudden rise of tea's status in social life in the Ming Dynasty related closely to the living places— Jiangsu and Zhejiang of the founders of the Ming Dynasty, these places were homeland of tea and key area of the early-stage regime of the Ming Dynasty.

Asked my wife making tea with fresh Xuanzhou tea leaves as tribute and offered it in the ancestral temple on March 14th, 1364.[12]

如事生，事亡如事存。'其言可法。今制宗庙祭器，只依常时所用者。"于是造酒壶盃盏之属。皆拟平时之所用。又置、筥、帷幔之属，皆象其平生焉。[14]

洪武二年冬十月壬戌朔，庚午敕葬开平忠武王常遇春于钟山之阴，赐明器九十事。[15]

这其中包括锡造金裹的茶盅、茶盏各一。

洪武二年二月己巳，奉先殿成，……上及皇太子诸王二朝皇后率嫔妃日进膳羞，每月朔荐新，正月用韭荠生菜鸭子鸡子；二月水芹薹、蒌蒿、子鹅；三月新茶、笋、鲤鱼；四月杏、梅、樱桃、黄瓜、雊雉；五月来禽桃李、茄子；六月莲子西瓜甜瓜；七月梨枣菱芡葡萄……[16]

丁亥，礼部又定祭祀礼器用瓷器，笾豆之属用竹器。

洪武九年二月丙戌定诸王公主岁供之数：

亲王岁支米五万石，钞二万五千贯，锦四十匹，纻丝三百匹，沙罗各一百匹，绢五百匹，冬夏布各一千匹，绵二千两，盐二百引，茶一千斤，马匹草料

Ceremonials Chamberlain offered seasonal items first as sacrifice in ancestral temple then into the royal court in May 1369 (2nd year of the Hongwu Reign).[13]

The production of ritual wares using 8880 taels of gold for ancestral temple was completed onJune 24th, 1369 (2nd year of the Hongwu Reign). Each ancestral temple would be allotted one jar, one alms bowl, two cups, two Jues, four Zhis, ten trays, four bags, two spoons, two teapots, two bells, one incense burner, one incense container, two vases, and two candlesticks. Originally, the emperor wanted to have gold ritual wares made for the ancestral temple. He instructed the protocol officer: "Ritual affair should be in line with human feelings, not have to be consistent with the ancient prototype. We usually use ancient Bian and Dou sacrifice containers in our worship ceremony. Taizu Emperor of the Song Dynasty once said: 'Our ancestors also did not know this.' Confucius said, 'Treat the deceased ones as they are alive.' We should learn from this. Ritual wares we made are all needed to be used in daily worshiping among which some are daily utensils of the deceased ones when they are alive."[14]

Chang Yuchun, King of Kaiping, was buried at the southern side of the Mount Zhongshan with 90 pieces of ritual wares including one cup and one calyx made of tin and wrapped with gold bestowed by the emperor on October 1st, 1369 (2nd year of the Hongwu Reign).[15]

The construction of the Fengxian palace was completed on February 6th, 1369 (2nd year of the Hongwu Reign)... Taizu Emperor, crown prince, kings, queens and concubines offered sacrifice every day in the Fengxian palace, changed sacrifice into new ones at the beginning of every month such as leeks, shepherd's purse, lettuce, duck and chicken in January; water celery, artemisia selengensis and goose in February; fresh tea,

第三章 明代茶道文化整体评价与茶画创作
Chapter Three Overall Evaluation of Tea Culture and Tea-Related Painting Creation of the Ming Dynasty

229

月支五十匹……

靖江王岁赐米二万石，钞一万贯，与物比亲王减半。

公主未受封，纻丝一十匹……

亲王子男未受封，纻丝一十匹……

亲王子女未受封，比男减半……

男已受封郡王，每岁支拨米六千石，钞二千八百贯……盐五十引，茶三百斤……"[17]

在这些供给中，亲王年得茶1000斤，受封亲王儿子得茶300斤。而公主、郡主和未受封的亲王儿子则没有资格获得茶芽。

从用茶荐祭太庙、纪念先祖的规定看，这无疑将茶的政治文化地位提到空前的高度。

而在廷殿当中向忠臣良将赐茶，则是最高荣誉。朱元璋对群臣说：

"朕闻太上为圣，其次为贤，其次为君子。宋景濂事朕十九年，未尝有一言之伪，诮一人之短，始终无二，非止君子，抑可谓贤矣。"每燕见，必设坐命茶，每旦必令侍膳，往复咨询，常夜分乃罢。[18]

bamboo shoots, and carp in March; apricot, plum, cherry, cucumber, pig, and pheasant in April; birds, peach, plum, and eggplant in May; Lotus seed,watermelon, and honey melon in June; pear, Chinese date, water chestnut, gorgon fruit, and grapes in July...[16]

The Ministry of Rites ordered on February 24th, 1369 (2nd year of the Hongwu Reign) that porcelain, bamboo Bian and Dou sacrifice containers should be used in worship ceremony.

The annual supply for the kings and the princesses was fixed on February 23rd, 1376 (9th year of the Hongwu Reign):

The annual supply for the prince is fifty thousand Dans (a unit of dry measure for grain=l00Sheng) of rice, twenty-five thousand Guans (a unit of coin) of money, forty bolts of brocade, three hundred bolts of ramie, one hundred bolts of gauze and satin each, five hundred bolts of tough silk, one thousand bolts of Winter and Summer cloth each, two thousand taels of cotton, two hundred Yins (a unit of length= 1.8m) of salt, one thousand Jins (a unit of weight = 0.5kg) of tea and forage for fifty horse each month...

The annual supply for the Jingjiang King is twenty thousand Dans of rice, ten thousand Guans of money which is half of that of the prince.

The annual supply for the untitled princess is ten bolts of ramie.

The annual supply for the untitled son of prince is fourteen bolts of ramie.

The annual supply for the untitled daughter of prince is seven bolts of ramie.

The annual supply for the title-granted duke is sixty thousand Dans of rice, two thousand eight hundred Guans of money... fifty Yins of salt and three hundred Jins of tea...[17]

Among these supplies, as to tea, one thousand Jins is for

四、明代以江南文士为主的茶人集团成为茶道文化的主体力量

台湾学者吴智和认为明代中晚期形成了以苏州为中心的茶人集团。这个集团的共识是出于自然性的结集，它所隐含的意义有多重，如性情、志趣、嗜好、交游、才学、名望等都是取得集团核心人物认可的依据。在当时中国商品经济最为突出的苏州文苑，形成以才学（诗文书画）为取向，而不单视阶层的优质风尚，才真正是社会文化进展的动因之一。[19] 他又将此集团内纯粹饮茶者分为隐逸茶人和寄怀茶人两个类型，把时间放在明中叶景泰至隆庆（1450—1572）之间，并试图从中探查明代茶人的精神高度。他认为饮茶进入文人集团的性灵世界有助于提升生活文化。性灵文化因饮茶（茶会）地点、主题不同而分为静态、动态、玩赏、创作和茶艺五大类型或五个方面。杜濬等茶人认为茶有四妙：湛、幽、灵、远；四用：澡、美、敏、导。在两宋茶人基础上，明代茶人试

prince, three hundred Jins for the title-granted son of prince, no bud tea for princess and untitled sons of prince.

Offering tea as a sacrifice in the ancestral temple showed undoubtedly the unprecedented high political and cultural status of tea in the Ming Dynasty.

Granting tea by the emperor to the loyal officials and excellent generals was regarded as the supreme honor in the royal court. Zhu Yuanzhang, Taizu Emperor of the Ming Dynasty, proclaimed, "I heard that the most admirable person is saint, then virtuous one and noble one. Song Jinglian worked for me for 19 years, he had never said single fake word, never talked of others' shortcomings, had been always loyal to the royal court. He is a real virtuous man." At all banquets when Song Jinglian was there, Taizu Emperor always arranged a seat and ordered tea for him, prepared breakfast every morning for him and discussed questions with him until late in the evening.[18]

D. South-of-Yangtze-River-Region Scholars Playing Main Role in Tea Culture of Ming Dynasty

Wu Zhihe, a Taiwan scholar, thought that a tea-man group centered in Suzhou was formed in the middle-late Ming Dynasty. The consensus of the group was spontaneity on the basis of personality, interests, hobbies, talent and learning, making friends and fame of which all were the criterion for getting recognition from core figures of the tea-man group. The fashion of taking one's talent and learning (mainly poems, calligraphy and painting) instead of social class as the sole standard in the literary world of Suzhou which was China's most advanced commodity economy at that time was one of the impetuses for social and cultural progress.[19] Wu Zhihe also divided the pure tea drinkers in this group lived between Jingtai

第三章 明代茶道文化整体评价与茶画创作
Chapter Three Overall Evaluation of Tea Culture and Tea-Related Painting Creation of the Ming Dynasty

231

图从情、悄、灵、素、韵、奇、绮、倩等方面深化茶道文化内涵。吴智和特别强调饮茶的性灵生活虽然无益于世道人心，但对人性适志的安身立命有莫大助益。[20] 应该说，苏州吴地的结社自有传承，应该与我们前述唐式茶会和中唐时期湖州茶文化圈影响有关。"张简，字仲简，吴县人……洪武三年荐修元史……当元季，浙东、西士大夫以文墨相尚，每岁必联诗社，聘一二文章钜公主之，四方名士毕至，宴赏穷日夜，诗胜者，辄有厚赠。临川饶介为元淮南行省参政，豪于诗，自号醉樵，尝大集诸名士赋《醉樵歌》。（张）简诗第一，赠黄金一饼，高启次之，得白金三斤；杨基又次之，犹赠一镒。"[21] "顾德辉，字仲瑛，昆山人，家世素封，轻材结客，豪宕自喜。年三十，始折节读书，购古书、名画、彝器、秘玩，筑别业于茜泾西，曰玉山佳处，晨夕与客置酒赋诗其中。"河东张翥、会稽杨维桢等"咸主其家"，"园池亭树之盛，图史之富暨饩馆声伎，并冠绝一时。"[22] 此处杨维桢应是《煮茶梦记》的作者。

to Longqing period (1430−1572) in the mid−Ming Dynasty into two types: hermit and tea lover, and tried to probe the spiritual world of tea man in the Ming Dynasty. He thought drinking tea in the literati group would help to promote quality of one's daily life. The spirit of tea culture contains five aspects— static, dynamic, joyful, creative, and tea art depending on different site and theme. Du Jun and other tea men believed that there were four wonderful features about tea such as crystal clear, serene, bright, and remote, and also four functions— purify spirit, beautify face, change mood, and guide mind. Tea men attempted to deepen the connotation of tea culture of the Ming Dynasty from the aspects of mood, serenity, intelligence, simplicity, rhyme, rarity, beauty and smile on the basis of ideas of tea men in Northern and Southern Song dynasties. Wu Zhihe particularly emphasized that drinking tea did no good for the way of the world and the heart of human being, but did benefit human nature.[20] We should say that Suzhou had its own traditions of establishing an association influenced by that of tea party of the Tang Dynasty and Huzhou tea culture in the mid−Tang Dynasty. "Zhang Jian, styled himself Zhongjian, was born in Wuxian County... was recommended to study history of the Yuan Dynasty in 3rd year of the Hongwu Reign... at the beginning of each season, scholars and officials from eastern and western Zhejiang gathered together and appreciated each others' paintings and calligraphies, poetry clubs were contacted to invite poets coming and recommending one or two poems to all for making comment during the day−and−night banquet every year, and the winners of poem would be awarded prizes. Rao Jie from Linchuan styled himself Zuiqiao (means "drunken woodman"), political consultant of Huainan Province in the Yuan Dynasty, loved writing and appreciating poem very much, once composed a prose 'Drunken Woodman Song'.

可以说，以苏州为核心区域的茶人集团代表了当时中国文人茶的最高水平。但我们发现这个集团多侧重于与"茶艺"相关的活动，从某种程度而言，如吴智和所言，无益于世道人心的引导和改变。

作为中国茶道文化发源地的湖州，工诗擅画者代不乏人。"王蒙，字叔明，湖州人，赵孟頫之甥也。敏于文，不尚榘度。工画山水，兼善人物……洪武初，知泰安州事。"[23] 新喻人梁寅所谓的"清、慎、勤，居官三字符也"[24]。在精神实质上与茶道文化有很大程度的契合。他们不仅注重自我修养，更对事关国计民生的茶事直接提出自己的看法。例如鄞人傅恕洪武二年诣阙陈治道十二策，就有"罢榷盐，停榷茶"[25]。

对茶学或对中国茶道文化做出准确判断的应该是朱权《茶谱》。《茶谱》成文于宣德三年（1429）至正统十三年（1448）间。《茶谱》云："予尝举白眼而望青天，汲清泉而烹活火，自谓与天语以扩心志之大，符水火以副内炼之功。得非游心于茶灶，又将有俾于修养

Zhang Jian won the first prize for poem and got one golden cake as reward, Gao Qi won the second prize and got three kilograms of platinum, Yang Ji got one Yi (an ancient unit of weight=20 or 24 Liang) for the third prize."[21] Gu Dehui, styled himself Zhongying, was born in Kunshan County into a rich family, loved to make friends with his unstrained character and broad-mindedness. He changed his usual behavior and began to learn knowledge at the age of 30, also bought ancient books, famous paintings, bronze wares and other antiques, built another villa called "Beautiful Place at Mount Jade" west of Xijing in which he drank spirits and composed poem with his guests day and night. Zhang Zhu from Shanxi and Yang Weizhen from Kuaiji were among his guests. "The landscape of garden, pool, pavilion together with huge amount of books and the condition of housing, eating and drinking provided in his villa were the best for a time."[22] Yang Weizhen mentioned in here should be the author of *Boiling Tea Dream*.

So to speak, the tea-man group centered in Suzhou represented the highest level of Chinese literati tea. But we found out that the tea-man group focused mostly on the tea art activities, which were not conducive to the world and the people's guidance and change, as Wu Zhihe said.

There has always been no lack of such person who was good at poetry and painting in Huzhou which was the birthplace of Chinese tea culture. "Wang Meng, styled himself Shuming and was born in Huzhou, was Zhao Mengfu's nephew and sensitive to literature but with ordinary level while his landscape and figure painting was of very high level... In the early period of Hongwu reign, Wang Meng was appointed as country magistrate in Tai'an."[23] In Liang Yin's opinion, "Being an official, one should be incorrupt, cautious, and diligent."[24] It fitted the spiritual essence of tea culture which emphasized the

第三章 明代茶道文化整体评价与茶画创作
Chapter Three Overall Evaluation of Tea Culture and Tea-Related Painting Creation of the Ming Dynasty

233

之道，其惟清哉"，"茶之为物，可以助诗兴而云山顿色，可以敷睡魔而天地忘形倍清谈而万象惊寒，茶之功大焉"。这是对《茶经》和《大观茶论》的直接继承。他在陆羽《茶经·十之图》之外，又对主、客献茶、接茶时的语言进行规范，而且在喝完茶，"话久情长后，礼陈再三，遂出琴棋，陈笔砚。或赓歌，或鼓琴，或弈棋，寄形物外，与世相忘。斯则知茶之为物 可谓神矣！"而张源著《茶录》以名词解说形式对采茶、饮茶及香、色、味等提出判断标准，以为茶中三昧。最后为"茶道：造时精，藏时燥，泡时洁。精、燥、洁，茶道尽矣"。虽然只是短短一段话，但我们认为是对中国茶道文化中形而下的高度概括，而不单是技术性问题，应该说是标准问题。茶道文化形而下的诸多标准确立，正是《茶录》的价值所在。因而陈继儒在撰写《茶话》时，将这三条标准放置于全文的最前面。

包括《茶谱》《茶话》《茶录》及许次纾《茶疏》，明代茶学著作五十余部，占中国古代茶书一半。与茶道文化发展的时代特点有关：

self-improvement. Some one such as Fu Shu from Ningbo of Zhejiang Province directly put forward his view in 2nd year of the Hongwu Reign about national economy and people's livelihood "Stopped the taxation and monopoly of tea and salt".[25]

It should be Zhu Quan who made an accurate judgment on Chinese tea science or Chinese tea culture by writing *Tea Manual* which was finished between 3rd year of the Xuande Reign (1429) and 13th year of the Zhengtong Reign (1448). He wrote, "I once gazed at the blue sky, drew water from clear spring and boiled it on lively fire because I believed one can enlarge heart by talking to the heaven, self cultivate by coinciding with water and fire, getting rid of worry and trying to be in peace and quiet will be an ideal state." And "Tea can do things such as stimulates poetic mood which darken the color of clouds and mountains, tames compulsive desire to sleep while leaving the earth and the heaven behind, creates warm atmosphere to idle talk while cooling the surroundings." Zhu Quan's views in this article was a direct inheritance of *The Classic of Tea* by Lu Yu and *Discussion on Tea* by Zhao Ji. Zhu Quan added besides of corresponding description in "Chart as Tenth Section" of *The Classic of Tea*. Some language criterion on offering tea by host and receiving tea by guest. "Guqin (seven-stringed plucked musical instrument), the game of Go, writing brush, and ink stone were displayed after tea drinking, then all the guests began to recite poem, play music or the game of Go after the long talk representing friendly feeling and the ritual behavior showing respect for the host until the moment everyone forgot the outside world." Zhang Yuan put forward his own standard on tea-leaves picking and tea drinking in *Tea Records* by means of nouns explanation. He concluded, "Essence of tea culture contains— processing

茶道形式从饼团点注向叶茶冲泡过渡，新的名品争奇斗艳，茶器由贵建盏向重紫砂发展。因此专论紫砂茶器的成书于 1640 年前后的周高起所著《阳羡茗壶系》和介绍吴中以立夏所采为贵的介绍岕茶的成书于 1608 年左右的熊明遇所著《罗岕茶记》为专题性著作。此外有不少是属于对历代茶事进行总结的著作，也有专门收集历代茶诗文的成书于 1611 年前后的喻政的《茶集》等。另外，明代茶学专家对水特别注重，田艺蘅《煮泉小品》（撰于 1554 年）为代表，继有徐献忠《水品》等。以陆羽《茶经》为基础，对新的泉水、品茗用水的新发现进行归纳总结。

总之，明代是中国茶事的持续发展时期，也是中国茶学繁荣时期。在借茶以张扬个性、借茶以明志、借茶以养廉示清高方面，达到历史的新高度。但是在茶道思想发展方面并没有达到陆羽所期望的高度。虽然"榷茶易马"成为明初既定国策，但茶道文化偏重茶艺和茶器、新茶的发现，没有组织化、系统化地发展，只停留在怡神养性，兼以琴棋书画

delicately, preserving dryly, and brewing cleanly." Although the conclusion was short, it presented a high level overview on Chinese tea culture at the same time provided a standard for Chinese tea culture rather than a technical problem. The establishing of standards for Chinese tea culture was made in *Tea Records*, this was just the value of this book. As a result, Chen Jiru put these three standards at the very beginning of his book *Tea Talking*.

Books on tea in the Ming Dynasty were over 50 including *Tea Manual, Tea Talking, Tea Records*, and *Tea Principle* by Xu Cishu, occupying half of ancient Chinese tea books. This phenomenon related to the era characteristics of tea-culture's development such as tea ceremony was in a period of transition from the form of tea cake point-to-point brewing to leaf tea wholly brewing, new famous tea brands appeared constantly, purple clay tea wares were valued more than Jian-kiln tea wares. *Purple Clay Teapot* specifically on purple clay tea ware written by Zhou Gaoqi was finished around 1640 and *Notes of Luokom Tea* by Xiong Mingyu on luokom tea finished around 1608. In addition, there were many books served as summaries of tea activities of previous dynasties, also a special collection of past dynasties tea poetry. *Tea Poem Collection* by Yu Zheng finished around 1611, books of the Ming Dynasty on water used for making tea were *On Spring Water for Making tea*by Tian Yiheng written in 1554, *On Tea-making Water Quality* by Xu Xianzhong summarizing his new findings on tea-making water based on *The Classic of Tea* by Lu Yu, etc.

In a word, the Ming Dynasty was a period for sustainable development of Chinese tea culture also a booming period of Chinese tea science. Ming-Dynasty tea culture reached a new height in history in the aspect of making individual character widely known, expressing one's ambition, cultivating honesty,

第三章 明代茶道文化整体评价与茶画创作
Chapter Three Overall Evaluation of Tea Culture and Tea-Related Painting Creation of the Ming Dynasty

235

的方面，还没有达到以茶导生气、借茶以治国的高度。

五、以"明四家"为代表的画坛与茶苑天然一体

在阅读《明史》时我们发现，与前朝元宋唐隋时期的传记不同，《明史》专列《文苑传》与《儒林传》分别纪传人物，而且《文苑传》的传主多为江南士人，以江都、长洲、浙江、松江（今上海）占绝大比例。这应当与这里的经济发达、诗画昌盛有密切关系，同时这里也是以瀹茶道所宗绿茶的重要产地。品茗论艺成为江南士人日常生活的重要组成部分。后来史家称道的画坛"明四家"实际也是德艺俱佳的大茶人。茶香茶韵弥漫于他们生活的方方面面。《文苑传》共四卷，入传诗书画士人凡126人。这126人仅仅是当时画坛中声名较为显赫的一部分，想必更多的士子茶人，并无缘入传。例如，在整个明代，创作出涉及茶事最多的诗画作品的明四家之一的仇英，在《明史·文苑传》里，几无只言片语。还有丁云鹏、陈洪绶等在画坛茶画史颇有影响，但亦未入传。钱毂等只是在

purity and dignity through tea drinking, and related activities. Regrettably, this height was not as high as Lu Yu expected. Tea culture aiming only at refreshing character developed from the aspect of ceremony accompanied by music playing, the game of Go or Chinese chess playing, calligraphy and painting creation, tea wares appreciating and finding of new tea types rather than systematization although "Taxation and Monopoly of Tea and Tea-horse Interchange Trade" turned into national policy in the early Ming Dynasty, so the height of tea culture was far from guiding the management of state affairs.

E. Art Circle Represented by Four Ming-Dynasty Master Painters and Tea Culture Circle Integrating into One

We found out while reading *History of Ming Dynasty* that it differed from biographies of Sui, Tang, Song, and Yuan dynasties by having "Literary Circle" and "Scholars' Circle" for historical figures. Most of literati recorded in "Literary Circle" section of *History of Ming Dynasty* were from south-of-the-Yangtze-River regions for example Jiangdu, Changzhou, Zhejiang and Songjiang (today's Shanghai) due to this area as a major green-tea producer, prospered poetry & painting and its advanced economy. Drinking tea and discussing tea art was an important part of daily life of literati of south-of-the-Yangtze-River region. Four Ming-Dynasty master painters—Shen Zhou, Wen Zhengming, Tang Yin, and Qiu Ying respected highly by historians of later dynasties were actually great tea men with virtue and skill. "Literary Circle" section of *History of Ming Dynasty* contained four volumes with 126 poets, calligraphers, painters of which these 126 people were only famous figures in art circle at that time, more scholars and tea men were not lucky enough to be recorded. For example,

《文徵明传》中，一句话提及而已。留下著名茶画的人物虽没有入传，但实际仍是茶道文化的拥趸者。同时，入传的大部分诗画名家，又没有茶画留下。但他们对品茗的热衷自然毫不逊色。

沈周（1427—1509），字启南，号石田、白石翁、玉田生、有居、竹居主人等。汉族，长洲（今江苏苏州）人。系明代中期文人画"吴派"的开创者，与文徵明、唐寅、仇英并称"明四家"。其与茶道文化有关的作品有《清明上河图》等。吴智和先生对茶人集团的阐述，已经使我们看到，从沈周到文徵明、唐寅，他们实际就是茶人集团的组成骨干，而沈周是早期盟主。

沈周一生家居读书，精勤书画创作。要书求画者"屡满户外"，"贩夫牧竖"向他求画，从不拒绝。甚至有人作他的赝品，求为题款，他也欣然应允。有曹太守其人，新屋落成欲图其楹庑，搜罗画家，沈周亦在其中，隶往谒之。沈周曰："毋惊老母，且夕往画不敢后。"客人颇不平曰："太守不知先生，何贱先生于此？谒贵游可勿往。"

Qiu Ying who was one of the four Ming-Dynasty master painters and created most of the poems and paintings on tea culture in the Ming Dynasty was mentioned only in a few words; Ding Yunpeng and Chen Hongshou who were known well in art circle and history of Chinese painting were both not included; Qian Gu and other important artists were only mentioned in one sentence in "The Life of Wen Zhengming"; other famous tea-culture painters who were also tea lovers are not included. Meanwhile most famous poets and painters recorded in "Literary Circle" section did not have their tea-culture paintings left behind but their enthusiasm for tea drinking was not inferior in any respect.

Shen Zhou (1427-1509), whose style name was Qinan, assumed nameswere Shitian, Baishiweng, Yutiansheng, Youju, Zhujuzhuren (means "a host of bamboo house"), etc. He was born in Changzhou (today's Suzhou in Jiangsu Province) with Han nationality. He was the pathfinder of "Wu School" of literati painting of Mid-Ming Dynasty also one of the four master painters of the Ming Dynasty sharing the same reputation with other three painter— Wen Zhengming, Tang Yin, and Qiu Ying. One of his paintings "Riverside Scene at Qingming Festival" related to the tea culture. We can see very clearly from Mr. Wu Zhihe's description that four master painters of the Ming Dynasty actually were backbones of tea-man group centered in Suzhou in the Middle-Late Ming Dynasty. Shen Zhou was the early leader.

Shen Zhou spent all his life time reading books, creating absorbedly and diligently works of calligraphy and painting at home. The door of his home had always been jammed with people from all walks of life who loved his works and begged to get one piece of them. He never refused request of any one. Some even imitated him and asked him to sign his name

第三章 明代茶道文化整体评价与茶画创作
Chapter Three Overall Evaluation of Tea Culture and Tea-Related Painting Creation of the Ming Dynasty

237

沈周答曰："往役义也，岂有贱哉？谒而求免，乃贱耳。"沈周的书画流传很广，真伪混杂，较难分辨。文徵明因此称他为飘然世外的"神仙中人"。

沈周法师黄庭坚，绘画造诣尤深，出入于宋、元各家，主要继承董源、巨然以及元四家黄公望、王蒙、吴镇的水墨浅绛体系。又参以南宋李、刘、马、夏劲健的笔墨，融会贯通，刚柔并用，形成粗笔水墨的新风格，自成一家。为吴门画派的领袖，也是茶人集团的最早盟主。

文徵明（1470—1559），长洲（今江苏苏州市吴县）人，祖籍为明代时期的湖广行省衡州府衡山县（今衡阳市衡东县）。明代中期最著名的画家、大书法家，号"衡山居士"，世称"文衡山"，官至翰林待诏，私谥贞献先生。"吴门画派"创始人之一。家世武弁，自祖父起始以文显，父文林曾任温州永嘉知县。自幼习经籍诗文，喜爱书画，文师吴宽，书法学李应祯，是中国书学中欧体书法的杰出代表。绘画宗沈周。少时即享才名，"与祝允明、唐寅、徐祯卿辈相切劘，名日益著"，

on their fake paintings, he did it joyfully. Once there was a prefecture chief surnamed Cao who wanted to decorate his newly built house with paintings of famous painters including Shen Zhou. The prefecture chief sent a servant to ask Shen Zhou for one painting. Shen Zhou said, "Do not disturb my mother, I will do it for your master in one day." The servant felt it was unfair to Shen Zhou, he said, "My master does not know your fame, how could he belittled you like this? You actually do not have to fulfill his desire." Shen Zhou answered, "I take this as my duty, no slight feeling of being belittled. If I am asked by others but refuse them thus I degrade myself with my behavior." Shen Zhou's calligraphy and painting spread very widely, fake pieces existed together with genuine ones making the differentiating difficult. Wen Zhengming therefore called him "immortal man" floating above the earthly world.

Shen Zhou learned from the skill of Huang Tingjian and other calligraphers and painters of the Song and Yuan dynasties especially the painting technique of ink and light purple system commonly adopted by Dong Yuan and Juran of the Song Dynasty, Huang Gongwang, Wang Meng and Wu Zhen of the Yuan Dynasty, also learn technique of ink-painting of Liu Songnian, Li Tang, Ma Yuan, Xia Gui of the Southern Song Dynasty, finally he achieved especially deep painting attainments. He started "Wu School" by creating a new style of broad writing-brush ink-and-wash painting with soft and hard lines on the basis of digesting comprehensive nutrition and adding his own feature of character. Shen Zhou was the soul character of "Wu School", also early leader of tea-man group centered in Suzhou in the Middle-Late Ming Dynasty.

Wen Zhengming (1470-1559) was born in Changzhou

并称"吴中四才子"。然在科举道路上却很坎坷，从弘治乙卯（1495）至嘉靖壬午（1522），十次落第，直至54岁因工部尚书李充嗣的推荐，以贡生进京为翰林院待诏。盛名于书坛画苑，登门求其书画者踏破门槛，引同僚的嫉妒和排挤。目睹官场腐败，抑郁不爽，57岁辞归出京，回苏州老家。晚年更是人书俱老，画登仙境，"文笔遍天下"，"海宇钦慕，缣素山积"。"其为人和而介"，年近90岁时，还孜孜不辍，为人书墓志铭，尚未写完，"便置笔端坐而逝"。

文徵明虽然曾腾达于官场，独步于诗坛画苑，但豁达而不失严谨，不沾名利。早在16岁父亲文林去世时，就坚决不接受温州吏民的千金赙礼。巡抚俞谏、宁王朱宸濠的厚礼都被拒绝。不接受曾被父亲提携的张璁、杨一清的拉拢。"四方乞诗文书画者，接踵于道，而富贵人不易得片楮。尤不肯与王府及中人（宦官），曰：'此法所禁也。'周、徽诸王以宝玩为赠，不启封而还之。……吴中自吴宽、王鏊以文章领袖馆阁，一时名士沈

(today's Wuxian County of Suzhou City in Jiangsu Province), ancestral home located in Hengshan County of Hengzhou Prefecture in Huguang Province (today's Hengdong County in Hengyang City) of the Ming Dynasty, whose style name was Hengshan Practitioner and Wen Hengshan, self posthumous name Mr. Zhenxian. He was one of the most famous painters and calligraphers in the Mid-Ming Dynasty, one of the creators of "Wu School" as well. He worked as an academician in the National Academy in the Ming Dynasty government. He was raised in a civil official family with his grandfather being good at writing and his father being a county magistrate in Yongjia of Wenzhou. He started to learn Confucian classics, poetry and liked painting and calligraphy since childhood. He learned writing from Wu Kuan; calligraphy from Li Yingzhen who was a prominent representative of calligraphy type of Ouyang Xun in Chinese calligraphy; painting from Shen Zhou. He shared a certain reputation for his talent at young age as one of "the Four Geniuses of Wuzhong" of which other three are Zhu Yunming, Tang Yin, and Xu Zhenqing. But he experienced failure ten times in the imperial examination from 1495 (52nd year of the Hongzhi Reign) to 1522 (19th year of the Jiajing Reign), was finally recommended by Li Chongsi— minister of Labor Ministry at age of 54 to work as an academician in the National Academy. He enjoyed a high reputation in circles of both calligraphy and painting, a huge amount of people came to him asking for one piece of his calligraphy or painting causing envy then squeezing out action of his colleagues. He quitted his job, left Beijing and returned his hometown in Suzhou at age of 57 after felt depressed by seeing with his own eyes the corruption in official circles. Wen Zhengming's calligraphy and painting gradually tended to become superb in his later years and were spread widely within the country. He

第三章 明代茶道文化整体评价与茶画创作
Chapter Three Overall Evaluation of Tea Culture and Tea-Related Painting Creation of the Ming Dynasty

239

周、祝允明辈与并驰骋，文风极盛。徵明及蔡羽、黄省曾、袁袠、皇甫沖兄弟稍后出。而徵明主风雅数十年，与之游者王宠、陆师道、陈道复、王穀祥、彭年、周天球、钱穀之属，亦皆以词翰名于世。"[26]他的茶事画有《真赏斋图》（上海博物馆）、《茶具十咏图》（北京故宫博物院）、《孝感图》（辽宁省博物馆）、《林榭煎茶图》（天津艺术博物馆）、《品茶图》《浒溪草堂图卷》及《惠山茶会图卷》（北京故宫博物院）等。他是茶人社团的又一位领袖人物。

唐寅（1470—1524），字伯虎。出身商人家庭，父亲唐广德，母亲邱氏。自幼聪明伶俐，20余岁时家中连遭不幸，父母、妻子、妹妹相继去世，家道中衰，在好友祝允明的规劝下潜心读书。29岁参加应天府公试，得中第一名"解元"。30岁赴京会试，却受考场舞弊案牵连被迁谪为吏，"寅耻不就，归家益放浪。宁王宸濠厚币聘之，寅察其有异志，佯狂使酒，露其丑秽。宸濠不能堪，放还。筑室桃花坞，与客日般饮

treated others mildly and moderately, worked diligently nearly 90 year old, even wrote epitaphs for the public and died when he was writing.

Although Wen Zhenming once played the game well in officialdom, obtained unique fame in the circles of poetry and painting, he still insisted to be open-minded while rigorous, keep distant from the bad habits in the vanity fair. As early as he was 16 when his father Wen Lin died, he refused to accept money as gift from people of Wenzhou. Moreover, generous gifts from Yu Jian— provincial governor and Zhu Chenhao— prince of Ning were turned away; precious gifts from Prince Zhou and Prince Hui were returned without opening the wrappers; did not accept the luring of Zhang Cong and Yang Yiqing who were guided and supported by his father. People from different places jostled each other in a crowd coming to beg one piece of his work of poem, calligraphy and painting and usually left with satisfaction, but the rich ones went away with empty hands especially those from palace of princes and eunuch. He once said: "This is banned by law." Celebrities came out in succession in Wuzhong area such as Wu Kuan and Wang Ao as leaders for a period, then Shen Zhou and Zhu Yunming, later on Wen Zhengming, Cai Yu, Huang Xingzeng, Yuan Qiu and Huangfu Chong. Wen Zhengming had dominated the fashion of literary elegance for decades with whom those had learning activities were all famous literati including Wang Chong, Lu Shidao, Chen Daofu, Wang Guxiang, Peng Nian, Zhou Tianqiu and Qian Gu.[26] Wen Zhengming was another leader of tea-man group whose tea activity paintings were "My Friend's Private House" preserved in Shanghai Museum, "Tea Tasting" preserved in Beijing Palace Museum, "Filial Piety" preserved in Liaoning Provincial Museum, "Entertaining Friend with Tea" preserved in Tianjin

其中，年五十四而卒"[27]。他临终时写的绝笔云："生在阳间有散场，死归地府又何妨。阳间地府俱相似，只当飘流在异乡。"

其山水、人物、花鸟见长。随周臣学画，后师法李唐、刘松年等。其画山重岭复，峰高水长，而笔墨细秀，布局疏朗，风格秀逸清俊，给人心旷神怡之感。人物画多为仕女及历史人物。其《事茗图》是传世名画之一，现藏北京故宫博物院。

祝允明，字希哲，长洲人。生长枝指，故号枝山。弘治五年（1492）举于乡，久之不第，授广东兴宁知府。"五岁作径尺字，九岁能诗。稍长博览群集。文章有奇气，当筵疾书，思若涌泉。尤工书法，名动海内。"[28]是唐寅的好友兼诗画同侣。

仇英（约1498—1552），与"明四家"中前三人不同，仇英出身工匠，早年曾当过漆匠，为富人寺观彩绘栋宇，后专于画事，被文徵明赞之为"异才"，董其昌称赞"十洲为近代高手第一。"仇英年轻时以善画结识了许多当代名家，受文徵明、唐寅

Art Museum, "Judging Tea", "Thatched Cottage", "Tea Party at Foot of Mount Huishan" preserved in Beijing Palace Museum, etc.

Tang Yin (1470−1524), style name "Bohu", was born in a merchant family with his father named Tang Guangde and his mother Qiu. He had been clever since childhood. Misfortunes had befallen upon this family when he was over 20 years old, his parents, wife and young sister died successively. The family fortune declined and he began to devote himself to learning after being exhorted by good friend Zhu Yunming. Tang Yin took part in Yingtian Prefecture level imperial examination and won the first place at age of 29. He went to Beijing wishing to attend metropolitan examination at 30, but was involved in the examination fraudulent case and exiled to be a petty official. He did not take the position regarding it as a shame, being unstrained more than ever when he was home. Zhu Chenhao— prince of Ning hired him with high salary. Tang Yin noticed that prince of Ning had the intention of rebellion, he thus pretended to be mad, behaved willfully while being drunken. The prince of Ning could not bear his ugly performance and released him. Tang Yin built a peach blossom castle for himself, drank all day long with his guests in it, died at age 54.[27] His last words went like this: "There will befinally a goodbye in this world. It doesn't matter if you die and go to hell. The two worlds are similar, death is nothing but wandering in a strange land."

Tang Yin was an expert in landscape, figures, flowers and birds painting. Learned painting technique as a student of Zhou Chen, then from imitating Li Tang, Liu Songnian, etc. He painted overlapped high mountains and water−flowing hills with thin and delicate lines, sparse composition, free and cute style giving us a relaxed and happy feeling. He took court ladies

第三章 明代茶道文化整体评价与茶画创作
Chapter Three Overall Evaluation of Tea Culture and Tea−Related Painting Creation of the Ming Dynasty

241

所器重，拜周臣门下学画，并曾在著名鉴藏家项元汴、周六观家中目睹了大量古画名作，大量临摹精品为创作积累了扎实基础。他既工设色，又善水墨、白描，能运用多种笔法表现不同对象，或圆转流美，或劲丽艳爽，《明画录》谓其："发翠豪金，丝丹缕素，精丽艳逸，无惭古人。"画山水以青绿为多，细润明丽而风骨劲峭，董其昌称其"赵伯驹后身，即文（徵明）、沈（周）亦未尽其法"。偶作花鸟，亦清丽有逸致。茶画有《东林图》《山间试泉图》等，临摹宋朝人的画作，几乎可以乱真。擅写人物、山水、车船、楼阁等，尤长仕女图，例如《清明上河图》。其茶事画《清明上河图》《赤壁图》与《松溪论画图》均藏辽宁省博物馆。

董其昌，字玄宰，松江华亭人。举万历十七年（1589）进士，改庶吉士。礼部侍郎田一儁以教习卒官，董其昌请假，千里扶柩归葬。天启五年（1625）拜南京礼部尚书。崇祯九年（1636）卒，年83岁。书法师承宋代米芾，画兼宋元诸

and historical figures as the theme for his figure painting. One of his most famous paintings "Self-entertainment by Drinking Tea" is now preserved in Beijing Palace Museum.

Zhu Yunming, style name Xizhe and also Zhishan because there were six fingers on his right hand, was born in Changzhou. He passed the Provincial Civil Service Examination in 1492 (5th year of the Hongzhi Reign), later on failed in further examinations for a long time, got an position as Xingning Prefecture magistrate in Guangdong. He "wrote calligraphy at age of 5, poem at 9; read extensively when he was older; his article had anunusual character, once even finished one piece of calligraphy with flows of ideas on a feast; hence obtained a nation-wide fame for his calligraphy."[28] Zhu Yunming was one of Tang Yin's good friends also companion of artistic creation.

Qiu Ying (1498-1552) had been a lacquerer who painted pictures on walls & ceilings in Buddhist temples and Taoist temples owned by the rich in his early years, later began to focus on artistic creation of painting. He was regarded by Wen Zhengming as "extraordinary talent" and by Dong Qichang as "No.1 master of art at his times". Qiu Ying met a number of contemporary famous painters with his superb painting technique when he was young. He learned painting as a student of Zhou Chen recommended by Wen Zhengming and Tang Yin who thought highly of him. He had a chance to view and imitated a large quantity of ancient masterpieces of painting at homes of Xiang Yuanbian and Zhou Liuguan who were eminent connoisseurs and collectors. Qiu Ying was good at using color, ink and brush stroke, line drawing and other techniques to embody different sorts of image especially court lady bearing taste of beauty flowing with roundel, spiral, powerful or colorful composition just as what was said

家之长，自成风格，"潇洒生动，非人力所及也"，"尺素短扎，流布人间，争购宝之。精于品题。收藏家得片语只字以为重。性和易，通禅理，萧闲吐纳，终日无俗语"。[29]

入《文苑传》的还有徐渭等。另外留下茶事画的还有丁云鹏、杜琼、钱榖等人，及许多作品艺术水品很高的无名氏。

纵观有明一代的茶事画（包括文人读书、听泉、观瀑、赏花、抚琴等），我们发现除像宋徽宗《文会图》那样描绘茶事场景的画作，以及文徵明《惠山茶会图卷》、谢环《杏园雅集图卷》等少数作品外，多是一人、二人或三人的品茗、观画、论书等情景。这表明茶人陈继儒所谓"品茶，一人得神，二人得趣，三人得味，七八人是名'施茶'"[30]的价值和审美判断实际是明代茶人共识。这也是明代茶道文化重在格物致知修身养性的高度所在。

六、名茶辈出，茶肆遍布都市城镇

由于明太祖等明代统治者将茶

about his art in Ming Painting Record. "Perfect, colorful and graceful images formed by thin, refined, exquisite and elegant lines in various colors such as green, golden, red or pure black are as good as those ancient master pieces." Qiu Ying's landscape painting mainly in green and bluish green had its own style with delicateness and soaked softness, brightness and beauty, vigor and stern strength. Dong Qichang once said that Qiu Ying was a reincarnation of Zhao Boju of Southern Song Dynasty, even Wen Zhengming and Shen Zhou could not be comparable to him. Qiu Ying painted birds and flowers occasionally bearing feature of elegance and beauty also carefree mood. The copies of Song-Dynasty paintings including that of figure especially court lady, landscape such as "Riverside Scene at Qingming Festival", cart, boat, pavilion, etc. done by him look almost the same as the genuine ones. Qiu Ying's tea activity painting are "For Mr. Donglin" and "Pine Pavilion by a Spring", and the following three pieces preserved now in Liaoning Provincial Museum—"Riverside Scene at Qingming Festival", "The Red Cliff", and "Discussing Painting under Pine Tree by Creek".

Dong Qichang, style name Xuanzai, was born in Huating of Songjiang. He passed the Civil Palace Examination in 17th year of the Wanli Reign (1589), then obtained the title of Hanlin Bachelor in Department of Study of National Academy. Dong Qichang asked a leave from his work to escort coffin of his teacher Tian Yijun, who was also a Hanlin Bachelor and assistant minister of the Ministry of Rites back to Datian County of Fujian several thousands miles away in the south. Dong Qichang began to served as a minister of Ministry of Rites of Nanjing in 5th year of the Tianqi Reign. Died in 9th year of the Chongzhen Reign at age 83. He learned calligraphy and related technique from Mi Fu of the Song Dynasty,

第三章 明代茶道文化整体评价与茶画创作
Chapter Three Overall Evaluation of Tea Culture and Tea-Related Painting Creation of the Ming Dynasty

243

视为一种战略资源，因而以陕南金州、汉中和四川等地的"边茶"地位极为突出。主要作为外销茶"边茶"的陕茶与川茶发挥重要作用。"设茶马司于秦、洮、河、雅诸州，自碉门、黎、雅抵朵甘、乌斯藏，行茶之地五千余里。山后归德诸州，西方诸部落，无不以马售者"，不久"于是永宁、成都、筠连皆设茶局矣"，"洪武三十年改设秦州茶马司于西宁，敕右军都督曰：'近者私茶出境，互市者少，马日贵而茶日贱，启番人玩侮之心。檄秦、蜀二府，发都司官军于松潘、碉门、黎、雅、河州、临洮及入西番关口外，巡禁私茶之出境者。'"[31] 为了适应少数民族的饮食习惯，明初开始出现"乌茶"与"黑茶"。

而其他所产茶叶统称为"腹茶"，除上供京师之外，多用于满足内地消费。"其上供茶，天下贡额四千有奇，福建建宁所贡最为上品，有探春、先春、次春、紫笋及荐新等号。旧皆采而碾之，压以银板，为大小龙团。太祖以其劳民，罢造，惟令采茶芽以进，复上供户五百家。凡贡茶，第按额以供，不具载。"[32] 除了从供应对象和制

inherited painting skills of famous painters of Song and Yuan dynasties. He established his own style: "Vivid, natural and unrestrained of which the height is hard for human being to reach." There were a lot of people visited him every day for getting one piece of his works, His notes and letters circulated widely in the society, people liked them very much and tried their best to purchase and preserved as treasures. Collectors felt lucky enough for getting even one piece of his calligraphy or painting. Dong Qichang was mild and easy-going, a Zen doctrine master, and had a refined style conversation all day longat home .[29]

Xu Wei were also included into "Literary Circle" of *History of Ming Dynasty*. Those whose tea-activity paintings survived were Ding Yunpeng, Du Qiong, Qian Gu and many anonymous artists with very high level of art.

It is not difficult for us to find out that besides of painting "Literati Tea Party" by Huizong Emperor of the Song Dynasty. "Tea Party at Mount Huishan" by Wen Zhengming and "Gathering in Apricot Garden" by Xie Huan, the themes of most of other tea-activity paintings of the Ming Dynasty were tasting tea, listening to waterfall or spring, watching flower, playing musical instrument, observing painting or discussing book by one, two or three persons after we taking an overall look at all of the Ming-Dynasty works of this type. It shows that the value and aesthetic judgment of tea founded by Chen Jiru. "As to tasting tea, ding it alone, one can understand the essence of value of tea; doing it by two persons, both can share an interesting time; doing it by three people, the taste of tea can be told. If seven or eight people do it together, it is just tea drinking."[30] It was widely accepted in the society. It embodies the height of the tea culture of the Ming dynasty whose core connotation is attaining knowledge through investigating things

作方法来进行分类外，随着散茶的普及，内地饮茶新品层出不穷。到明代中晚期，各地名茶各有高下，进入南京的名茶就有："如吴门之虎丘、天池，岕之庙后、明月峡，宜兴之青叶、雀舌、蜂翅，越之龙井、顾渚、日铸、天台，六安之先春，松萝之上方、秋露白，闽之武夷、宝庆之贡茶，岁不乏至。"[33] 同样《儒林外史》第二十四回，对南京茶坊亦有文学化描述："大街小巷，合共起来，大小酒楼有六七百座，茶社有一千余处。不论你走到一个僻巷里面，总有一个地方悬着灯笼卖茶，插着时鲜花朵，烹着上好的雨水，茶社里坐满了吃茶的人。"而张岱（1597—1680）的《斗茶檄》庄谐有趣，使我们对明代茶点有所了解："瓜子炒豆，何须瑞草桥边；橘柚查梨，出自仲山园内。八功德水，无过甘滑香洁清凉；七家常事，不管柴米油盐酱醋。一日何可少此，子猷竹庶可齐名；七碗吃不得了，卢仝茶不算知味。"一壶挥尘，用畅清淡；半榻焚香，共期白醉。

文学作品对茶肆多有描述，使我们得以领略茶语茶肆的情形：

in nature and cultivating morality constantly in one's whole life.

F. Famous Tea Types Appearing in Large Numbers and Tea Houses Scattering in Cities and Towns

Tea produced in the border areas of Jinzhou and Hanzhong in southern Shaanxi and Sichuan and had been the main export product shared a very special status in the Ming Dynasty because the rulers of Ming took tea as a kind of strategic resource. Tea and Horse Bureaus were set up in prefectures of Qin, Tao, He, Yazhu graranteeing the tea transportation in the five−thousand−Li area from Diaomen, Liya to Duogan and Wusizang. Those tribes in the Western Region sold their horses to get tea. Soon Tea Bureaus were established in Yongning, Chengdu, Yun and Lian. Tea and Horse Bureau in Qinzhou Prefecture was moved to Xining for management convenience in 1397 (30th Year of the Hongwu Reign), the army chief said that private tea had been exported, fewer tea trades were made in the market pushing the tea price down and horse price up at the same time aroused the insulting intention of neighbouring tribes. Tea Bureaus in Qin and Shu were closed and military troops were dispatched to Songpan, Diaomen, Liya, Hezhou, Lintao and strategic passes in order to inspect and forbid private tea export.[31] Redish brown tea and black tea were processed in order to adapt to the dietary habit of western minorities.

Other types of tea were called "Interior Product" of which some were transported to the capital, others were consumed in other places in the country. As for tea tribute to the royal court, more than four thousand catties of tea were graded among which Jianning tea from Fujian was of the highest quality including the following types Tanchun, Xianchun, Cichun, Zisun, Jianxin, etc. Taizu Emperor abolished the

第三章 明代茶道文化整体评价与茶画创作
Chapter Three Overall Evaluation of Tea Culture and Tea−Related Painting Creation of the Ming Dynasty

245

《水底鱼儿》

〔末上〕

开设茶坊。声名满四方。

煎茶得法。非咱胡调谎。

官员来往。招接日夜忙。

卢仝陆羽。也来此处尝。

也来此处尝……

《好姐姐》此茶十分细美。

看烹来过如陆羽。一泉二泉。试

尝君自知。休轻觑。路逢侠客须

呈剑。不是才人不献诗。

……好茶且请张员外。醉倒

王公旧酒垆。

……〔外〕多谢茶。三钱银

子在此。

〔末〕多谢了。[34]

写范仲淹在茶坊与店小二对
白了解下情的情景。从其对白来
看，茶坊伙计对茶的历史知识和
茶叶品鉴有相当的水准。另外，
我们应该明白，即使同样属于散
茶，也存在釜中煎煮和盏中冲泡
的两种茶道形式，而"团茶"的
瀹注形式也没有立即退出历史舞
台。至于民间的传统饮茶方式传
承更为持久，例如湖南与江西的
"擂茶"就有很强的生命力。

最后，作为具有世界影响的明

"Coiled−Dragon−and−Phoenix Pie" tea made by pressing with silver plate since the late Tang and early Song dynasties and took bud−tea instead as a tribute to the royal court, so five hundred families were fixed as royal tribute tea producers who should offer their products according to the assigned amount without taking the corvee.[32] New types of tea emerged successively in the country. In the middle and late Ming Dynasty, famous teas were in huge amount, what could be found in the market of Nanjing were Huqiu, Tianchi from Suzhou; Miaohou, Mingyuexia from valley; Qingye, Queshe, Fengchi from Yixing; Longjing, Guzhu, Rizhu, Tiantai from Yangzhou; Xianchun from Liu'an; Shangfang, Qiulubai from Songluo; Wuyi and Baoqing tea from Fujian as royal tribute.[33] [quoted from "Tea Grades" as volume 9 of *Wishes from My Guests* by Gu Qiyuan (1565−1628)] The literary description on tea house in Nanjing can be found in the 24th chapter of *Unofficial History of the Academic Circles* by Wu Jingzi. "There are more than six or seven hundred restaurants of different size, over one thousand tea houses in big streets and small lanes. You will always find a place with hanging lanterns, in−season flowers, boiling clean rain and many tea lovers in any side lane you walk up to." Humorous and interesting writings on tea drinking and pastries of the Ming Dynasty can be found in "Tea Competition Announcement" by Zhang Dai (1597−1684). "There is no need to buy melon seeds and fried beans by the Ruicao Bridge; orange, grapefruit and pear are from garden on Zhongshan Hill. Eight−merits water tastes sweet, clean and cool; seven homely things of the daily necessities are fuel, rice, cooking oil and salt. I can not live one single day without tea just like Wang Huizhi could not do without bamboo in his life; Lu Tong did not know the taste of tea until he drank seven cups."

朝，既往的历史学家注意到了郑和下西洋对世界的影响。对明代在茶叶的传播和茶道文化的普及方面的贡献，似乎较少涉及。一个明显的事实是，正是在明代中期，也就是日本茶道正式形成阶段："16世纪这一百年，是日本茶道史上最重要、最光辉的一百年。"[35]日本茶道的开拓者村田珠光和日本茶道集大成者千利休都相继出现在这个时代。对于日本茶文化为什么越过三个世纪，忽然在16世纪蓬勃发展起来的原因，至今没人给出合理回答。"终明之世，通倭之禁甚严，闾巷小民，至指倭相詈骂，甚以噤其小儿女云。"[36]这表明，日本与明朝关系不甚友好，但因相互博弈而交往的深度、广度，却是前朝无以为比的。倭寇掳掠明州、台州、福州、泉州、温州等地民众无数，更是对人口和文明的直接掠夺。

小结

综上所述，我们发现，明代统治者十分重视茶叶的经济战略作用和外交博弈功能，制定"榷茶易马"国策，无疑达到唐宋两朝未有的程度。在祭祀先祖等礼

Literary works on tea house provide us a vivid reference for getting to know the situation of tea activities and tea-related expressions:

Underwater Fish

(Tea House Owner) I ran a tea house for years and had been famous far and wide; I am not self-boasting in here, I cooked tea in the proper way; Officials came and went constantly, I have been busy day and night; Lu Tong and Lu Yu once even came here to taste my tea; came here to taste my tea...

Good Sister

This tea is very good and tasty, the tea cooking technique is better than that of Lu Yu; one can find high-quality water only by tasting the water from one spring to another, please do not look down upon this effort; show treasured sword only to the excellent swordsman when met on the road, should not offer poetry to the one who is not a gifted scholar.

...Good tea and here is Zhang Yuanwai (Ministry Counsellor). Any one will surely be drunk after drinking my tea.

[Zhang Yuanwai] Thanks for your tea, here is three silver coins as tea fee.

[Tea House Owner] Thanks a lot.[34]

Aforementioned description shows the dialogue between Fan Zhongyan and the tea house owner (namely the waiter). It is obvious from the dialogue that the tea house owner has quite some knowledge on the history of tea and tea tasting. In addition, we should know that there are two kinds of tea ceremonies for making tea with loose leaves: brewing tea in the cauldron and in the calyx. As for the traditional way of tea drinking, it has been inherited for a longer period with a very strong vitality such as "Lei Tea" from Hunan and Jiangxi provinces.

第三章 明代茶道文化整体评价与茶画创作
Chapter Three Overall Evaluation of Tea Culture and Tea-Related Painting Creation of the Ming Dynasty

247

仪上奉献茶汤，放置茶器，无疑提高了饮茶的政治地位。茶叶经济和茶叶消费也达到空前规模。茶肆林立遍布城市乡镇，从事茶事的人们对茶叶知识和茶文化的了解达到新的高度。

但是，以明代帝王为代表的政治家们，在对茶道文化的价值判断和精神意义的认识上，并没有达到唐宋帝王们的以茶引导清和的政治高度。中国茶道的形而下之艺则因为文人士大夫的热衷参与而空前发展，形而上的茶道思想则更多地表现为文人士大夫的个人修养。江南的茶人集团使茶道文化呈现出比前代更为繁荣的局面。新的茶道形式出现，新茶品如雨后春笋，新的茶器也不断成为宫廷和文人们的追求，瓷器、紫砂茶器呈现出异彩纷呈的景象；茶学著作也表现出丰硕的态势。但是我们不能不看到，明人由于浮躁自利的整体环境的影响，他们把陆羽《茶经》、赵佶《大观茶论》、蔡襄《茶录》所涵盖的以茶治国、以茶修身养性的中国茶道内涵消解在品茶鉴水的过分追求之中。《朱权》对茶道思

Finally, historians have noticed the worldwide influence of Zheng He's voyage to the western Pacific and Indian ocean in the Ming Dynasty, but few books or documents mentioned the contribution of the Ming Dynasty in spreading of tea and popularization of tea culture. In fact, Japanese tea ceremony was formally formed in the middle Ming Dynasty. "The one hundred year of the 16th Century is the most important and glorious period in the history of Japanese tea ceremony."[35] The pioneers of Japanese tea ceremony Murata Pearl and Japanese tea ceremony master Senno Rikyu both lived in this period. As for why the Japanese tea culture suddenly flourished in the 16th century across three centuries, no one has ever given a reasonable answer. "At the end of the Ming Dynasty, the ban for aiding the Japanese side was very strict, ordinary people cursed each other in name of the Japanese, even scared the kids to keep silent with mentioning Japanese people.[36] This suggests that Japan's relation with the Ming Dynasty was not very friendly, but the contact between two sides due to the national competition was much wider and deeper than the former dynasties. The Japanese side plundered countless people from Mingzhou, Taizhou, Fuzhou, Quanzhou, Wenzhou, and other places which was equivalent to a direct plunder of population and civilization.

Summary

In conclusion, we find out that the rulers of the Ming Dynasty attached great importance to the tea economy's strategic role and its function in the diplomatic game playing, then formulated the national policy "Taxation and Monopoly of Tea and Tea-horse Interchange Trade" and executed it undoubtedly to the degree Tang and Song dynasties had not achieved. At the ceremonies of worshiping the ancestors, tea

想把握较深，但过于简略。在茶学看似繁荣、茶叶以空前规模走向国外市场的情况下，中国茶道并没有在唐宋基础上实现突破，这是历史的后视镜，时时提醒我们，中国茶道文化的复兴之路，尚有一段艰难的上坡路要走。

虽然如此，我们从画坛"明四家"及董其昌等文苑诗坛的众多巨匠身上，又看到了资质刚健朴素清廉的苏轼笔下"叶嘉"的茶人风骨。这批茶人在诗、书、画、乐、文等方面达到空前高度，显得伟岸而又丰满，是明代茶道文化的真正主体，由于自身的努力，他们基本实现经济独立，既淡出官场，又与金钱能使鬼推磨的滚滚红尘保持相当的距离，茶人的独立与自尊达到历史新高度。不像陆羽需要颜真卿与皎然的援助，也不像苏轼，毕生难以脱离官场羁绊。他们敢于神闲气定地拒绝官场诱惑，可以将送上门来的金银珠宝原封不动地淡然奉还，并生活得游刃有余，逍遥自在。在茂林修竹之下，在白云苍松之间，在瀑布流泉之畔，文火慢煎，品天地之灵气，感自然之神妙，是

wares were placed, and tea were dedicated as sacrifice surely increasing the political position of tea drinking. The tea economy and tea consumption also reached an unprecedented scale. Tea houses scattered in cities and towns, the knowledge of tea and tea culture of those engaged in tea production, tea consumption and tea activities reached a new height.

The judgment of Ming-Dynasty politicians represented by the emperors on the value of tea culture and its spiritual significance had not reach the height of that of the Tang and Song dynasties emperors. Chinese tea ceremony developed unprecedentedly because of literati's enthusiastic participation while super-organic tea art contained mainly literati's personal accomplishment. The Tea-man Group based on south-of-Yangtze-River-region scholars brought the prosperity of tea culture for the Ming Dynasty embodied by the appearance of new forms of tea ceremony, novel tea products, new tea wares including porcelain and purple clay pieces and abundant tea-related literary and artistic works. We have realized that people of the Ming Dynasty digested the core connotation of Chinese tea culture— governing the country by tea and cultivating one's morality with tea expressed in the books such as *The Classic of Tea* by Lu Yu, *Discussion on Tea* by Zhao Ji, and *Tea Records* by Cai Xiang, by means of tasting tea and grading tea-making water due to the influence of the overall restless-and-self-interest-orientedenvironment. Zhu Quan presented his deeper understanding on tea culture in *Tea Manual* but his description was too brief. Chinese tea culture could not see its breakthrough in the Ming Dynasty although it looked prosperous and tea industry connected with international market. It reminds us that we still have a hard and long way to go to revitalize Chinese tea culture.

Nevertheless, we have seen tea-man's character— vigorous,

第三章 明代茶道文化整体评价与茶画创作
Chapter Three Overall Evaluation of Tea Culture and Tea-Related Painting Creation of the Ming Dynasty

249

明代文士茶的特征，是生活，是艺术，也是宗教。如许美妙的茶事画把人们带进那个遥远但曾经真实存在的品茗环境中，人们或深或浅地走进明代茶人的精神世界。明代茶道不如唐宋厚重大气，但也不失清新隽永的滋味。

（原文载《楚雄师范学院学报》，2015年第5期，总第200期，应赵荣光教授之约为该刊特撰）

【注释】

[1]［美］崔瑞德、［美］牟复礼编：《剑桥中国明代史》，中国社会科学出版社，2006年，第352页。

[2]《大明太祖实录》卷二百一十二。

[3]《大明太祖实录》卷二百一十七。

[4]《大明太祖实录》卷二百四十五。

[5]《大明太祖实录》卷二百五十一。

[6]《大明太祖实录》卷二百五十二。

[7]《大明太祖实录》卷二百五十三。

[8]《大明太祖实录》卷二百五十四。

[9]《大明太祖实录》卷二百五十四。

[10]《大明太祖实录》卷二百七十九。

[11]《大明太祖实录》卷二百三十二。

[12]《大明太祖实录》卷二十二。

[13]《大明太祖实录》卷四十二。

[14]《大明太祖实录》卷四十三。

[15]《大明太祖实录》卷四十六。

[16]《大明太祖实录》卷六十一。

[17]《大明太祖实录》卷一百四。

simple, honest, and upright— summarized by Su Shi of the Song Dynasty were presented by master painters of the Ming Dynasty including Shen Chou, Wen Zhengming, Tang Yin, Qiu Ying, and Dong Qichang. These Ming−Dynasty tea men as the main force of the tea culture at that time reached an unprecedented height in terms of poetry, calligraphy, painting, music and literature. They also obtained basic economic independence through their efforts so as to be able to get out of officialdom meanwhile kept a certain distance with the money−oriented earthly world hence maintained their spiritual independence and self−respect unlike their predecessors such as Lu Yu who had been always helped financially by Yan Zhenqing and Jiaoran, and Su Shi who had to deal with official affairs in officialdom for his life time. They dared to refuse in tranquility the temptation from officialdom, were able to return the gold, silver, jewelry intact with indifference and live comfortably and at ease with their own incomes. Their life was filled with vitality, artistic and religious atmosphere composed of natural wonders and heavenly aura from thick forest, tall bamboo, cloud, pines, waterfall, stream, slow fire and brewed tea. Tea activity painting of the Ming Dynasty brings us into that remote yet once existed tea−tasting surroundings, more or less into the spiritual world of the Ming−Dynasty tea men. Tea ceremony of Ming Dynasty was not as profound and dignified as that of Tang and Song dynasties still with its own fresh and meaningful taste.

【 Notes 】

[1]Denis Twitchett, Frederick W. Mote, eds., *The Cambridge History of China's Ming Dynasty*, China Social Sciences Publishing House, 2006, p.352.

[2]*The Memoir of Taizu Emperor's Reign of the Ming Dynasty*, vol.212.

[18]《明史》卷一百二十八《列传十六·宋濂忠传》，中华书局，1974年，第3786—3787页。

[19] 吴智和：《明代茶人集团》，《明史研究》第三辑，第122页。

[20] 吴智和：《中明茶人集团的饮茶性灵生活》，《史学集刊》1992年第4期，第61页。

[21]《明史·文苑一·张简》，第7321页。

[22]《明史·文苑一·顾德辉》，第7325页。

[23]《明史·文苑一·王蒙》，第7333页。

[24]《明史·儒林·梁寅》，第7226页。

[25]《明史·文苑一·傅恕》，第7319页。

[26]《明史·文苑三·文徵明传》，第7361页。

[27]《明史·文苑二·唐寅传》，第7353页。

[28]《明史·文苑二·祝允明传》，第7352页。

[29]《明史·文苑四·董其昌传》，第7396页。

[30] 陈继儒：《岩栖幽事》载《眉公杂著》。

[31][32]《明史》卷八十《食货志四：盐法茶法》。

[33] 顾起元：《客座赘语》卷九《茶品》。

[34] 范受益：《寻亲记》第三十三出《惩恶》。

[35] 滕军：《日本茶道文化概论》，东方出版社，1992年，第37页。

[36]《明史·外国三·日本》。

其他参考书籍还有：

张廷玉《明史》，中华书局，1974年版。

清《四库全书》第1231至1290册。

《大明太祖实录》。

阮浩耕、沈冬梅、于良子：《中国古代茶叶全书》，浙江摄影出版社，1999年版。

陈文华：《中国茶文化学》，中国农业出版社，2006年版。

[3]*The Memoir of Taizu Emperor's Reign of the Ming Dynasty*, vol.217.

[4]*The Memoir of Taizu Emperor's Reign of the Ming Dynasty*, vol.245.

[5]*The Memoir of Taizu Emperor's Reign of the Ming Dynasty*, vol.251.

[6]*The Memoir of Taizu Emperor's Reign of the Ming Dynasty*, vol.252.

[7]*The Memoir of Taizu Emperor's Reign of the Ming Dynasty*, vol.253.

[8]*The Memoir of Taizu Emperor's Reign of the Ming Dynasty*, vol.254.

[9]*The Memoir of Taizu Emperor's Reign of the Ming Dynasty*, vol.254.

[10]*The Memoir of Taizu Emperor's Reign of the Ming Dynasty*, vol.279.

[11]*The Memoir of Taizu Emperor's Reign of the Ming Dynasty*, vol.232.

[12]*The Memoir of Taizu Emperor's Reign of the Ming Dynasty*, vol.22.

[13]*The Memoir of Taizu Emperor's Reign of the Ming Dynasty*, vol.42.

[14]*The Memoir of Taizu Emperor's Reign of the Ming Dynasty*, vol.43.

[15]*The Memoir of Taizu Emperor's Reign of the Ming Dynasty*, vol.46.

[16]*The Memoir of Taizu Emperor's Reign of the Ming Dynasty*, vol.61.

[17]*The Memoir of Taizu Emperor's Reign of the Ming Dynasty*, vol.140.

[18]"Biography of Song Lian", *History of Ming Dynasty*, Beijing: Zhonghua Book Company, vol.128, 1974, pp.3786−3787.

[19]Wu Zhihe, "Tea−man Group of Ming Dynasty", *Research on History of Ming Dynasty*, vol.3, p.122.

[20]Wu Zhihe, "Tea Drinking as Part of Spiritual Life of Tea−man Group in Middle Ming Dynasty", *Collected Papers of History Studies*, vol.4, 1992, p.61.

[21]"Biography of Zhang Jian, First Part of Literary World", *History of Ming Dynasty*, vol. 285, p.7321.

[22]"Biography of Ge Dehui, First Part of Literary World", *History of Ming Dynasty*, vol. 285, p.7325.

[23]"Biography of Wang Meng, First Part of Literary World", *History of Ming Dynasty*, p.7325.

[24]"Biography of Liang Yin, First Part of Scholars World", *History of Ming Dynasty*, p.7226.

[25]"Biography of Fu Shu, First Part of Literary World", *History of Ming Dynasty*, p.7391.

[26]"Biography of Wen Zhengming, Third Part of Literary World", *History of Ming Dynasty*, p.7361.

[27]"Biography of Tang Yin, Second Part of Literary World", *History of Ming Dynasty*, p.7353.

[28]"Biography of Zhu Yunming, Second Part of Literary World", *History*

第三章 明代茶道文化整体评价与茶画创作
Chapter Three Overall Evaluation of Tea Culture and Tea−Related Painting Creation of the Ming Dynasty

251

陈宝良：《明代社会生活史》，中国社会科学出版社，2004 年 3 月版。

王天有、高寿仙：《明史：一个多重性格的时代》（台湾）三民书局，2008 年初版。

伊水文：《明代衣食住行》，中华书局，2012 年版。

《农业考古》，茶文化专号。

滕军：《日本茶道文化概论》，1992 年版。

《中国美术全集》（明代绘画），上海人民美术出版社，1988 年版。

中南海画册编辑委员会：《中国传世名画》，西苑出版社，1998 年版。

台北《故宫博物院藏画》

国家文物局：《惠世天工》，中国书店，2012 年版。

of Ming Dynasty, p.7352.

[29]"Biography of Dong Qichang, Forth Part of Literary World", *History of Ming Dynasty*, p.7396.

[30] Chen Jiru, "The Miscellaneous Things", On *Trivial Matters*.

[31]"Tax Method on Salt and Tea, Forth Section of Monograph on Food and Currency", *History of Ming Dynasty*, vol.80.

[32]"Tax Method on Salt and Tea, Forth Section of Monograph on Food and Currency", *History of Ming Dynasty*, vol.80.

[33]Gu Qiyuan, "Tea", *Guest Message*, vol.9.

[34]Fan Shouyi, "Punishing Evils", *The Search of Loved One*, episode 33 of drama.

[35]Teng Jun, *Introduction to Japanese Tea Culture*, Beijing: Oriental Press, 1992, p.37.

[36] "Japan, Third Part of Foreign Countries", *History of Ming Dynasty*, vol.230.

图 158

明墓壁画·《男侍》（摹本）

　　明代墓葬壁画，人物高 75 厘米。1998 年河南省登封市芦店镇明墓出土。原址保存。

　　墓向 183°。位于墓室东壁。双线勾出边框，框内绘厢房一座。右侧绘一黑色炉灶，炉上置一浅盘，盘内放一小口垂腹瓶。炉左侧一男子，着红色圆领广袖长袍，腰束绿带，足穿软底黑靴，右手提一茶壶，边向炉边走，边回首张望。壁画左侧绘一曲腿方桌，桌面右侧黑漆盆内放四只石榴，黑漆方盘内覆置五只小碗，盘前放三只茶盏。桌右侧一男子，梳圆髻，着黄色领广袖袍，腰束红带，足蹬软底靴，躬身捧一黑漆盘，盘内盛一带钮盖茶壶。

第三章　明代茶道文化整体评价与茶画创作
Chapter Three Overall Evaluation of Tea Culture and Tea-Related Painting Creation of the Ming Dynasty

253

图 159

明·李时·《清赏图》（局部）·辽宁省博物馆藏品

原画高 39.2 厘米，宽 35.8 厘米。

作者生卒年代不详，少师广孝（1335—1419）有诗赠之。画面最上方，一人端坐，旁穿深衣的书童手捧茶叶罐，一客人看红边木架、一客人回望书童，另一书童捧持四幅字画前来，一穿浅蓝衣服的书童，正捧着黄布囊包裹的古琴，第四位书童正在端着水盆，另一客人正用之沐手，主人可能就是没有戴幞头的坐瓷墩座者，一客人背向右手捧一云头如意。桌上有茶杯、茶罐、铜鼎、香炉、插花。品茶、赏花、赏琴的聚会刚开始。

图 160

明·蓝瑛（1585—1664）**·《草堂卧云图卷》·北京故宫博物院藏品**

画高 34 厘米，宽 176 厘米。

三位文士围几床而坐，有茶杯，一侍童恭立，旁边一小屋，有人专门准备茶汤。

第三章　明代茶道文化整体评价与茶画创作
Chapter Three Overall Evaluation of Tea Culture and Tea-Related Painting Creation of the Ming Dynasty

255

图 161

明·杜琼（1396—1474）·《南村别墅十景图册》·上海博物馆藏品

　　原画高 33.5 厘米，宽 50.9 厘米。

　　取材于明初文学家陶宗仪《南村别墅十景咏》，是对作家所在的云间（今上海松江）泗泾
别墅环境的客观描写。雪夜，炉火，清茶，友情，艺术，文法，书法，古今多少事，伴随茶香飘。
陶宗仪《南邨诗集》之《秋怀次戴景仁韵》云："丛桂吐幽芳，清泉茶鼎洁。"（《四库全书》
第 1231 册，上海古籍出版社，第 1231—1280 页）

图 162

明·杜琼·《南村别墅十景图册之四阁杨楼》·上海博物馆藏品

　　读书，是茶人日课的重要内容之一，因而，林下水滨，深山匡庐，高阁幽轩等是茶人
们与先圣时贤对话的绝妙佳境，而茶也是随时携带的庄严法器。

第三章　明代茶道文化整体评价与茶画创作
Chapter Three Overall Evaluation of Tea Culture and Tea-Related Painting Creation of the Ming Dynasty

257

图 163

明·杜琼·《友松图卷》·北京故宫博物院藏品

　　两人品茗话古今，又有一友人至，书童负琴跟随。笔筒、直流紫砂壶
就在曲谱旁边。

杜琼·《友松图卷》（局部）

图 164

明·杜琼·《南湖草堂图轴》

　　此画成于成化四年（1468）作者73岁时。云光树影错落有致，意趣浑茫，人笔俱老，归于质朴。苍山古柏，有高泉飞流。一个读书人与山水相融走进哲学的殿堂，侧身历史的刀光剑影，或者已经与古人唱和，物我两忘。这也是茶致幽静的境界。

第三章　明代茶道文化整体评价与茶画创作
Chapter Three Overall Evaluation of Tea Culture and Tea-Related Painting Creation of the Ming Dynasty

259

图 165
明·郭纯（1370—1444）·《人物》（局部）·台北故宫博物院藏品

　　郭纯，初名文通，赐今名，遂以文通为字，号朴庵，浙江永嘉人。以精绘事，擢营缮所丞。洪熙元年（1425）陞阁门使。山水学盛懋，布置茂密。有言夏圭、马远者，辄斥之曰："是残山剩水，宋僻安之物也，何取焉？"纯充内廷供奉，宣宗尝面促之画，不即执笔，以死怵之，对曰："苦书乐画。宁死不能草草。"

　　二人叙语，谈古论今，邀明月相陪，茶香通天。

郭纯·《人物》（局部）

从主人手中羽扇来判断，时序当在仲夏，二人话旧，"夜后邀陪明月"。

第三章 明代茶道文化整体评价与茶画创作
Chapter Three Overall Evaluation of Tea Culture and Tea-Related Painting Creation of the Ming Dynasty

261

图 166

明·沈贞（1400—1482）·《竹炉山房图》·辽宁省博物
馆藏品

作者为画家沈周的伯父。画高 115.5 厘米，宽 35 厘米。

作者题记：成化辛卯初夏余游毗陵（今常州），遇竹炉山
房，得普照师留酌竹林深处，谈话间出素纸索画，余时薄醉，
挑灯戏作此图，以供清赏。南齐沈贞。

左上角有乾隆御题：阶下回回㵳惠泉，竹炉小叩赵州禅，
个中我亦曾清憩，为缅流风三百年。庚辰仲秋御题

所谓"三百年"，当从绘画当年算起，应是实写。画中是
二人对谈，有白瓷茶壶，却不见茶杯。

沈贞·《竹炉山房图》（局部）

第三章 明代茶道文化整体评价与茶画创作
Chapter Three Overall Evaluation of Tea Culture and Tea-Related Painting Creation of the Ming Dynasty

263

图 167

明·谢环·《杏园雅集图》·镇江市博物馆藏品

    绢本设色，高 37 厘米，宽 401 厘米。

    作者生卒不详，洪武成名，永乐入宫，宣宗赐锦衣卫千户待遇。画中有成套的盏托，有酒瓮，

也有食盒、红漆茶盒。

    《杏园雅集图》对我们理解雅集与中国茶道文化关系有启发意义。

谢环·《杏园雅集图》（局部）

秦末高士东园公唐秉、甪里先生、绮里季吴实与夏黄公崔广四人避乱隐居商山，人称"四皓"。汉相张良辞官隐居留坝，被称为神仙。东晋永和九年（353），右军及谢安、孙绰等41位众士大夫在春明景和、惠风和畅的会稽山阴的兰亭水边举行自古流传的修禊仪式，沐浴濯足，心与天契，志同地合，快然自得，曲水流觞，饮酒赋诗，为中国士人集会之大端。

汉唐进取，"宁为百夫长，胜作一书生"。历经"安史之乱"，士人精神生活趋于平淡。颜真卿刺湖之时（772—777）聚集了有名有姓的唐代文人士大夫60余人，常常品茗论诗，为唐代茶道文化形成奠定基础。唐会昌五年（845）三月二十四日九位退休老人白居易、胡杲、吉皎、郑据、刘真、卢真、张浑、李元爽、释如满相聚洛阳履道坊白居易居所欢聚"尚齿"之会，既醉且欢之际赋诗画画。

宋朝尚文，以孝治国。中华文化进入成熟昌盛阶段。王诜（1048—1104）自幼天资聪敏，好读书，过目不忘。及长，诸子百家诗文、书画外，琴棋笙乐、百工技艺亦无不通晓。除自行创作外，照旧搜集书画与其同道交换共赏，又在他的私第之东筑宝绘堂蓄其所藏书画，被视为"风流蕴藉，真有王榭家风"。据米芾《画史》所载，他曾收藏过的珍贵书画就有：一、顾恺之人物一卷，米芾以怀素帖换走；二、荆浩画山水，后送与米芾；三、苏轼作《竹木枯石图》；四、黄鉴《凤牡丹图》屏条六幅；五、韩干《夜照白马》一幅；六、王士元《渝村浦屿》雪景四幅，得自收藏家王巩（定国）处；七、梦休《雪竹》图一幅，易自蒋长源处；八、江南人画《小雪山图》二轴，疑为王维画；九、米芾画《小树图》一幅；十、易元吉《苇芦鸲鹆》一幅，借自米元章；十一、颜真卿法书一卷；十二、李公麟画《西园雅集图》一卷（内有苏东坡、黄山谷、张耒、秦观、米芾、王诜、李公麟等16人图像）。

南宋杭州南湖别业也是张镃联系社会、交游雅集的一个开放性空间。张镃家世显赫，甚有艺，又非常好客，喜游意

风雅，与人吟诵唱和，交游极为广泛，举凡当时朝野政要如史浩、萧燧、洪迈、周必大、姜特立、京镗、楼钥，道学大家如朱熹、陈傅良、彭龟年、陈亮、叶适、蔡幼学等文坛名家如陆游、杨万里、尤袤、范成大、辛弃疾、姜夔等，均与之有往来唱酬。（曾维刚、铁爱花：《园林别业与宋人休闲雅集和文学活动——以杭州张镃南湖别业为中心的考察》，《浙江学刊》2012年第5期）

此图卷描绘了明正统二年（1437）三月一日，内阁大臣杨士奇、杨荣、杨溥及画家等十人在杨荣的杏园聚会之情景。图卷后保留着当时雅集者手迹：杨士奇的《杏园雅集序》，杨士奇、杨荣、杨溥、王英、王直、钱习礼、周述、李时勉、陈循等九位朝臣。杨荣在《杏园雅集序》中这样描述："倚石屏坐者三人，其左，少傅庐陵杨公[杨士奇，时为内阁首辅、少傅（从一品）、兵部尚书（正二品）兼华盖殿大学士]，其右为荣[杨荣，时为荣禄大夫（从一品）、少傅（从一品）、工部尚书（正二品）兼谨身殿大学士]，左之次少詹事泰和王公[王直，时为少詹事（正四品）兼侍读学士]"，这些是画幅中最重要的一组人物。《杏园雅集》被《翰林记》作为馆阁雅集的典范记录下来："正统二年三月，馆阁诸人过杨荣所居杏园燕集，赋诗成卷，杨士奇序之，且绘为图，题曰：杏林雅集。预者三杨二王、钱习礼、李时勉、周述、陈循与锦衣卫千户谢廷循也。荣复题其后，人藏一本，亦洛社之余韵云。"（黄佐：《翰林记》卷二十，影印文渊阁四库全书，596册，台湾商务印书馆，1983）尹吉男认为这是以江西进士团为主体的较为隐蔽的官场活动形式。与会昌九老图有较大差异，也与李公麟《西园雅集》的实际情形有差异。（尹吉男：《杏园雅集与杏园雅集图》，《故宫博物院院刊》，2016年第1期）

古代文人士大夫的雅集活动是茶道文化的重要表现形式。虽然雅集的主题与组织者各不相同，但往往离不开煮水、品茶与赋诗唱和。而以饮茶为主题的活动当以文徵明《惠山茶会图》最有名。

图 168

明·沈周（1427—1509）·《桂花书屋图》·北京故宫博物院藏品

　　原画高 153 厘米，宽 35.2 厘米。

　　沈周为"明四家"之首，也是吴中茶人集团第一代盟主。从他始，明人画事茶事相互熏染。茶灶紧邻书屋，在桂花树下，在简朴而整洁的书房内，主人静坐，闲舒而从容，在吱吱的煮水声中，到达物我两忘的境界。

图 169

明·沈周·《山水轴》·台北故宫博物院藏品

在崇山峻岭之下，茂林修竹之中，小桥流水之畔，画家雨后独自追赶，茶会衍期，携茶灶前往。

约期为成化丙申（1476）四月二十九日，作者次日"雨后振孤策，迢遥追往踪"，为友人吉之作画并题诗，使他们知道未能如期聚会者的落寞心情。

第三章 明代茶道文化整体评价与茶画创作
Chapter Three Overall Evaluation of Tea Culture and Tea-Related Painting Creation of the Ming Dynasty

267

图 170

明·沈周·《焚香品茗图》

　　画中，在青铜器的旁边是青花瓷。画面最左面，侍女手捧高坐带盖茶碗，也是青花瓷，画面最上面围棋石方桌旁的书童手捧青花瓷，既像是杯，有可能是围棋罐。画中主人正在接取穿绿衣黄花上衣的侍者盘中的青花瓷香炉。他身后两仕女奉茶。另一仕女端包袱。两个书桌上有打开的册页、古玩。桌旁有插花（似红珊瑚）。

图 171

明·沈周·《拙修庵》（局部）·南京博物院藏品

　　原画高28.6厘米，宽33厘米。

　　土墙之内，草庵之下有书橱、茶床、茶炉。炉上有匜形提梁镬。茶床上有红托白（青）釉茶盏。提梁锡壶。主人在小方榻上结跏趺坐。草庵掩映在绿树红花之下。这也可能是东洋草庵茶道的早期形态吧。是文人"茶禅一味"的简单实践，谓之"拙修"。饮茶间与书房是明显的两个相连的空间。

图 172

明·吴伟（1459—1508）·《词林雅集图》（局部）

　　炉边两瓮是贮存泉水的专用器具。唐宋茶学没有涉及贮水器，但茶人对水质的要
求程度，绝不在茶叶质量之下。

第三章 明代茶道文化整体评价与茶画创作
Chapter Three Overall Evaluation of Tea Culture and Tea-Related Painting Creation of the Ming Dynasty

271

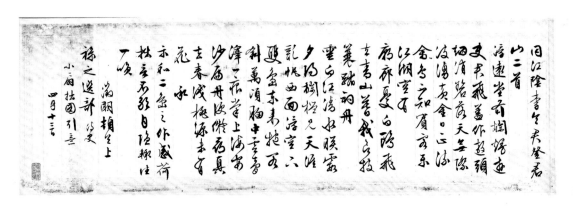

图 173

明·文徵明（1470—1559）·《林榭煎茶图》·天津博物馆藏品

　　画内落款和画后题记表明，画为友人禄之所作，时在初夏四月十三日，文徵明是继沈周以后茶人盟主，也是诗、书、画俱佳的文化大家。

文徵明·《林榭煎茶图》（局部）

第三章 明代茶道文化整体评价与茶画创作
Chapter Three Overall Evaluation of Tea Culture and Tea-Related Painting Creation of the Ming Dynasty

273

图 174

明·文徵明·《东园图卷》·北京故宫博物院藏品

　　竹林幽幽，古树苍苍，水波潋滟，评书、论画、较诗之时，佐以香茗，是明代士
人向往的生活。

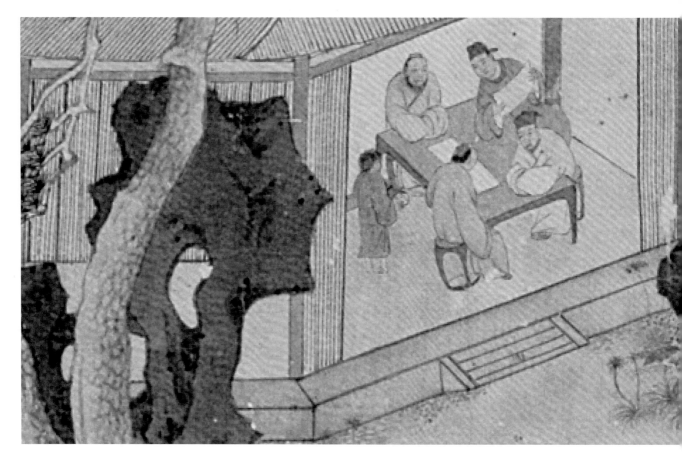

文徵明·《东园图卷》（局部）

第三章 明代茶道文化整体评价与茶画创作
Chapter Three Overall Evaluation of Tea Culture and Tea-Related Painting Creation of the Ming Dynasty

275

图 175-1

明·文徵明·《惠山茶会图》·北京故宫博物院藏品

　　此图绘于正德十三年（1518）二月十九日。据蔡羽序记，正德十三年二月十九日，文徵明与好友蔡羽、王守、王宠、汤珍等人至无锡惠山游览，品茗饮茶，吟诗唱和，十分相得，事后便创作。

　　在一片松林中有座茅亭泉井，诸人冶游其间，有的人围井而坐，展卷吟哦，有人散步林间，赏景交谈。其中一人观看书童烧水煮茶。此君，白色袍服，衣领褐色边，黑幞头。站在红色茶几前，双手拱揖。一童着深青色衣，好像在茶几下取东西，白色风炉以铁架筐固，可见炉中炭色灰黑，炉火红色，炉上为灰白色提梁壶，还像是锡铝制品。大口圆腹直而宽的提梁壶带圈足，长流平直，在另一个茶几上放置，很像黄铜制品。敞口深腹青铜器壶，无疑在这里做了贮水器。铅锡制品茶器有考古发掘品为证，四川崇州万家镇明代窖藏出土的明代永乐成化年间的铅锡执壶，专为煮茶而制作。（卢引科、李绪成、李开、刘雨茂、易立：《四川崇州万家镇明代窖藏》，《文物》2011年第7期）人物面相虽少肖像画特征，

大都雷同，动态、情致刻画却迥异，饶有生意，并传达出共通的闲适、文雅气质，反映了文人画家传神胜于写形的艺术宗旨。同时，青山绿树、苍松翠柏的幽雅环境，与文人士子的茶会活动相映衬，也营造出情景交融的诗意境界。

　　画前（左）由蔡羽所撰《惠山茶会序》云："渡江而润、金、焦，甘露胜；由润入句容，三茅山胜；由句容至毗陵，白氏园胜；由毗陵至无锡，惠麓山胜。余之之金陵。必经是傍。……尝与文徵明、中山唐子重、太原王履约、王履吉谋行。而诸君各有典守，又不敢舍己（已）业而以越人境。正德丙子之秋，长洲博士古闽郑先生掌教武进，居于毗陵。明年丁丑夏，吾师大教士太保靳公致政，居于润。又明年戊寅春，子重以父病将祷于茅山。履约兄弟以煮茶法，欲定水品于惠。其二月初九，余得往润之日，与诸友相见于虎丘，又辞以事，乃独与箭泾潘和甫挟舟去。子重亦与其徒汤子朋同载前后行。三宿达润。……二月十九日清明日……午造泉所，乃举王氏鼎立二泉厅下，七人者环亭坐。注泉于鼎。三沸而三啜之，识水品之高，仰古人之趣，各陶陶然不能去矣。于戏胜哉旬日之力耳。过者造，造者遍，

又获与友共矣！顾视畴昔何如哉？然世之熟视吾辈，则不能无疑以为无情于山水泉石，非知吾者也。以为有情于山水泉石非知吾辈也！诸君子稽高器也。为大朝和九鼎而未偶，姑适意于泉石，以陆羽为归将，以差时之乐红粉、奔权幸、角锱铢者耳。诸君屋漏则养德，群居则讲艺。清志虑开聪明则涤之以茗。游于丘，息于池用全吾神而高起于物，兹岂陆子所能至哉！固鲁点之趣也。会成赋诗，冠以序。正德十三年戊寅二月清明日林屋山人蔡羽撰。"

《序》中毗陵是今天常州，武进为其属县；焦，指润州今镇江焦山。如文徵明所记：每年谷雨之前，天池、惠山等茶区的饮茶活动进入繁盛阶段。各种规模各种形态的茶会活动相继举行。文人以茶为媒介的雅集只是其中的一种。从沈周、文徵明以来的江南苏州、常州、润州、金陵茶人集团，对文化艺术的推动具有不分阶层地域唯才艺是尚的积极意义。但在社会、经济、中外南北交流呈现加速发展的明代，江南茶人集团渐趋走向封闭与保守，唐宋时代以来以茶修身、藉茶悟道、以茶廓清社会风气从而达到以茶治国的高远意旨被消解，形而下的鉴水、品茗、赏画、听雨、浇花、高卧、翻经、焚香、洗砚等等

成为他们追逐的目标。就品茗活动中细微体验与划分，明代则达到前代未曾达到的高度。

徐献忠《水品》是以"源""清""甘""流""寒""品""杂说"为品水方向。佳泉宜茶，而用于品茶之水则更为讲究。明代许次纾《茶疏》认为："精茗蕴香，借水而发，无水不可与论茶也。"

文徵明在《谢宜兴吴大本所寄茶》："小印轻囊远寄遗，故人珍重手亲题。暖含烟雨开封润，翠展枪旗出焙齐。片月分明逢谏议，春风仿佛在荆溪。松根自汲山泉煮，一洗诗肠万斛泥。"《袁与之送新茶荐以荣夫新笋赋谢二君》："拣芽骈笋荐新泉，石鼎沙铛手自煎。"《煎茶诗赠履约》："嫩汤自候鱼生眼，新茗还夸翠展旗。"《次夜会茶于家兄处》："寒夜清谈思雪乳，小炉活火煮溪冰。"《煎茶》："竹符调水沙泉活，瓦鼎燃松翠鬣香。"这些诗文对我们理解明代茶事过程中的各种器具和品茗标准有极好的借鉴作用。陈继儒（1558—1639）《茶话》提出的"品茶一人得神，二人得趣，三人得味，七八人是名施茶"似乎过于个人化。但在提倡加强八小时外个人修养的今天，似乎还有学习体验的必要。

第三章 明代茶道文化整体评价与茶画创作
Chapter Three Overall Evaluation of Tea Culture and Tea-Related Painting Creation of the Ming Dynasty

277

惠麓烟中見名泉掛杖尋幽弱多碁未記
韓尚雲林昔有篇茶法人無飲水心清風澈修
竹山古得餘音

清明日同諸友宿湯亭
南北多歧法同共二有因廛雲和松里把酒愁
良辰客淡清江笛芹美 古驛春但教吾道在
錄食余萬貧
蔡羽

鄭博士官舍夜集
春帆慶梅柳連夕會江樓海內偏青眼天涯
易向頭夜遅北斗明月滿西州綠酒頻酌
渾鎖舊別愁
遊白氏園
石路去縈紆亭臺盡不如偕花扶繡柱借柳
暎春渠過客許金谷流觴賽夜除湖山雙
在眼鄭卜倖樵漁
宛轉山林趣都歸相國家鑿環竹塢水
接青霞落日情無已來鴻逸易除故園他夕
夢春色遠天涯
客散鳴聲鞭繁花春未休直愁白日不撓
鮮蘭丹悠酷王孫吟成鸚鵡洲昆陵風景
吳虛 踏青遊
雷別鄭博士
相憶情何限為歡不可窮離杯春水綠驛路
杏花紅軽檀歐外狐城細雨中江湖愁滿地
飄轉任萍蓬
惠山作
雨至青山晚曉春泉滑正流松雲合竹色珠雪
瀟龍湫品盥中冷下茶堪北苑技名賢雷脈
賞合向水經收
送別丹陽暮懐人茂苑遠東風食細雨
花朝舟中
漏花朝草樹迎船過煙雲停水消西津江路
攔堤今何夕雷船依春水坐人聚德
望亭丹中諸反夜集
星看歲月淪江漢雲霄梯羽翰轉蓬吾眾惜
愁殘酒杯乾

中山 湯珍

徑曲知何處突兀林坳辞易豬堪補心塵卽更洗閒趣
好評泊水經茶譜童子語山深豚游地伴杯悶取喚熱
青餐霞竹裡幽情付與 長伴暗谷泉生渥湮蘭繆停
雅亭畫樹危關漫撫慈雪金鼎內融得一壺春聚惟悴玉
川人坐清畫危關漫撫松庭屑哨品翠陰送路五日
舊雨江湖遠鴻漸重奎細吽浮梅珑香無幽徑湝芳井
韻龍吻春雲玉滅囊挽試新湯典吹適金鑪暖記留連
派花漲膩松風古澗 都是惜別行跫送客歸向幕
江目斷寫情題水葉山簪映一二 半斜清淺倒影洗
閒媛共留取生綃淨膚瀗冰泉亂峯鎮任紅塵一坪 羅山
右詞共題文衡山惠山茶會圖卅圖秀潤古雅士氣
盎然為衡山生平傑作倘令松雪見之六當斂手何況餘子
同治七年季冬中浣艮蕃居士識於過雲樓

图 175-2

明·文徵明·《惠山茶会图》·序文、诗文

这是在中国茶道文化史上具有重要意义的一次活动：茶人集团盟主文徵明邀约好友蔡羽、王守、王宠、汤珍、唐子重、潘机甫，相聚于惠山泉旁，主要内容是用清明前天池、惠山茶和王氏兄弟著名的茶鼎来鉴定惠山泉；活动后，不仅有画、

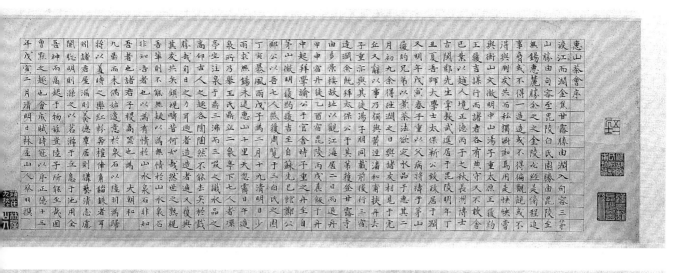

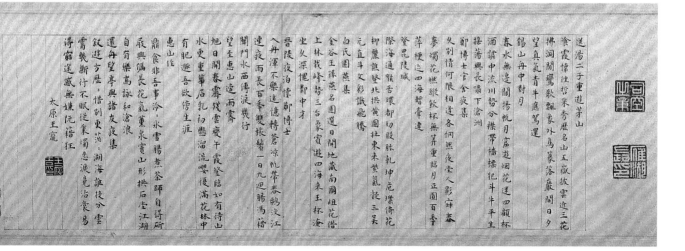

有序文、有诗歌创作、唱和，特别是蔡羽序文最后一段话，高度涵盖了茶人精神，从而升华了明代茶道文化高度：以陆羽为归将，对"乐红粉、奔权幸、角锱铢"的世风持坚决批判，划清界限的明朗态度，屋漏则养德、群居则讲艺，以佳茗清泉来清志虑、开聪明，目的在于"为大朝和九鼎！"这是明代中国茶道文化的最高境界！

第三章 明代茶道文化整体评价与茶画创作
Chapter Three Overall Evaluation of Tea Culture and Tea-Related Painting Creation of the Ming Dynasty

279

图 176

明 · 文徵明 · 《品茶图轴》 · 台北故宫博物院藏品

题款: "碧山深处绝纤尘, 面面轩窗对水开。谷雨乍过茶事好, 鼎汤初沸有朋来。嘉靖辛卯山中茶事方盛, 陆子傅过访, 遂汲泉煮而品之, 其一段佳话也! 徵明制。"可知, 时在 1531 年。钤白文朱印: "文徵明印", 朱文"乾隆""御赏之宝"等数印。

文徵明·《品茶图轴》（局部）

白衣者师，端坐书案正中，儒雅正定；背向者弟子，浅浅地坐在木凳一半，手轻轻地抚案小角，认真、谦恭地谛听为师说教。白瓷杯与鼓腹茶壶里的茶香弥漫书房！

第三章 明代茶道文化整体评价与茶画创作
Chapter Three Overall Evaluation of Tea Culture and Tea-Related Painting Creation of the Ming Dynasty

281

图 177

明·文徵明·《茂松听泉图》

　　两个年长的文友在茂盛的松林里，在潺潺的泉水边，谛听泉水上涌、流动的声音。这与烹茶过程中依靠声音判别水沸程度是一致的，是将人的情绪和思维与大自然融为一体，饮茶与否，茶浓茶淡，倒在其次。

图 178

明·文徵明·《茶具十咏图轴》·北京故宫博物院藏品

画高 136 厘米，宽 26.8 厘米。

此与前边《惠山茶会图》有联系，又有不同。还是天池虎丘茶事最盛的清明谷雨之际，友人每举行茶会。今年（1534）自己只能抱病独处一室，品试送来的两三种新茶，偶然想到唐代皮日休、陆龟蒙咏茶、咏器具的诗句，聊寄一时之兴，内容有茶坞、茶人、茶、茶具和著茶过程。

第三章 明代茶道文化整体评价与茶画创作
Chapter Three Overall Evaluation of Tea Culture and Tea-Related Painting Creation of the Ming Dynasty

283

图179

明·文徵明·《绝壑高闲图》

　　文徵明《暮春斋居即事》："经旬寡人事，踪迹小窗前。暝色连残雨，春寒宿野烟。茗杯眠起味，书卷静中缘。零落梅枝瘦，风吹更可怜。"这首诗是其晚年生活写照。另有《题虢国夫人夜游图》云："紫尘拂辔春融融，参差飞鞚骄如龙。锦鞯绣带簇妖丽，绛纱玳烛围香风。春风交花光属路，后骑雍容前却顾。中间一骑来逡巡，秀眉玉颊真天人。翠微垂鬓极称身，仿佛当年虢与秦。佳人绝代真难得，安得君王不为惑？岂知尤物祸之阶，不独倾城竟倾国。一时丧乱已足怜，后世方夸好颜色。晴窗展卷漫多情，百年青史自分明。莫言画史都无意，尺素还堪鉴兴废。"本诗隐括杜甫《丽人行》诗意，不仅介绍了画面和画中人物形态，同时还强调了画的讽刺作用。

　　《题画·山行》："高涧落寒泉，穷岩带疏树。山深无车马，独有幽人度。幽人何所从？白云最深处。出山不知遥，愿见人间路。"诗篇介绍了画中环境，高涧寒泉，穷岩疏树，这是林栖者的最佳地，也是历代文人当他们仕途偃蹇，便向往栖息山林，过着避世绝尘的生活。虽云消极，但还有一种与统治者不合作的反抗心态。

图 180

明·文徵明·《乔林煮茗图》·台北故宫博物院藏品

此图文徵明绘于 1526 年五月，作者 56 岁。

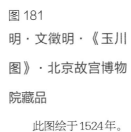

图 181

明·文徵明·《玉川图》·北京故宫博物院藏品

此图绘于 1524 年。

第三章 明代茶道文化整体评价与茶画创作
Chapter Three Overall Evaluation of Tea Culture and Tea-Related Painting Creation of the Ming Dynasty

285

图 182

明·文徵明·《影翠轩图轴》·台北故宫博物院藏品

　　可能由于为避免烟火熏燎，茶灶一般安置在书房之外。

文徵明·《影翠轩图轴》（局部）

第三章 明代茶道文化整体评价与茶画创作
Chapter Three Overall Evaluation of Tea Culture and Tea-Related Painting Creation of the Ming Dynasty

287

图 183

明·文徵明·《真赏斋图》·上海博物馆藏品

　　原画高 36 厘米，宽 107.8 厘米。作者 80 岁时的作品。

　　这幅画中又出现了宋元时常用的白盏红托，从侍者用食箸搅拌的情形看，表明是在煮散茶。煮好倒入茶壶，用盏托品饮。

"真赏斋" 是无锡收藏家、太学生华夏（字中甫）的书斋收藏室之名，文徵明应约作此画。修竹、茂松及太湖尽在画中，以煮水烹茶、谈诗论道为主题，足见茶事画是衡山先生的擅长作品。

文徵明·《真赏斋图》（局部）

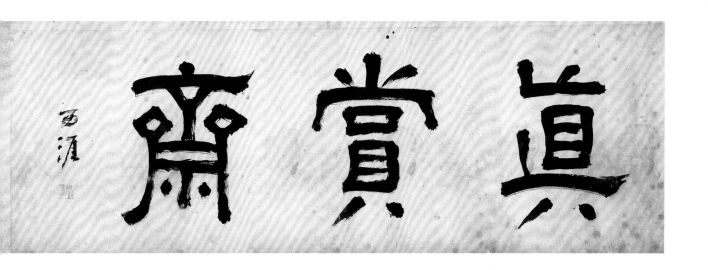

第三章 明代茶道文化整体评价与茶画创作
Chapter Three Overall Evaluation of Tea Culture and Tea-Related Painting Creation of the Ming Dynasty

289

图 184

明 · 文徵明 · 《寒山风雪图》

　　风紧雪急，夫人探窗，老翁围炉备茶、温酒。从风雪里归来，
一杯热茶在手，是身体的温暖，是心灵的慰藉。

文徵明·《寒山风雪图》（局部）

第三章 明代茶道文化整体评价与茶画创作
Chapter Three Overall Evaluation of Tea Culture and Tea-Related Painting Creation of the Ming Dynasty

291

图 185

明·丁云图·《罗汉册》

    原画高 30.9 厘米，宽 31.2 厘米。

    疑与丁云鹏（1547—1628）为同一人。有丁丑至戊寅图成之题，知画于
1577 至 1578 年之间。侍者左手所提为茶壶无疑。

丁云图·《罗汉册》（局部）

第三章 明代茶道文化整体评价与茶画创作
Chapter Three Overall Evaluation of Tea Culture and Tea-Related Painting Creation of the Ming Dynasty

293

日長何所事茗碗
自齎持料得南
窓下清風滿鬢
綠吳趨唐寅

图 186

明·唐寅（1470—1524）·《事茗图》·北京故宫博物院藏品

　　高方提梁壶可能是紫砂壶，直接瀹茶。客人不仅带书童携琴，还自带茶杯。足见其
真茶道中人也。

唐寅·《事茗图》（局部）

图 187

明·唐寅·《款鹤图》·上海博物馆藏品

原画高 29.6 厘米，宽 145 厘米。

文士款鹤，侍童为文士备茶！

图 188

明·唐寅·《煮茶图》·台北故宫博物院藏品

原画高 21.6 厘米，宽 96 厘米。

图 189

明·唐寅·《煎茶图》·台北故宫博物院藏品

原画高 106.9 厘米，宽 48.1 厘米。

第三章 明代茶道文化整体评价与茶画创作
Chapter Three Overall Evaluation of Tea Culture and Tea-Related Painting Creation of the Ming Dynasty

297

一宿團欒遊旅中殘詞聊以
識泥醉當時我作陶穀音
何必尊前面發紅　唐寅

图 190

明·唐寅·《陶穀赠
词图轴》·台北故宫
博物院藏品

陶穀（903—970）
为五代宋时彬州人，著
有《清异录》，对茶事
有简明扼要的叙述。鍑
中坐茶执壶，表明古人
也"温茶"，是陆羽"茶
宜趁热连饮"原则的新
表现形式。

畫棟珠簾煙水中落霞孤鶩渺
無蹤千年想見王南海曾借龍王
一陣風
　　圖
　　　晉昌唐寅為
　　德輔辞兄先生作詩意

图 191

明·唐寅·《落霞孤
鹜图》·上海博物馆
藏品

落霞之时，孤鹜远
飞，在高山之腰，落水
之滨，茶灶安闲无烟，
茶童相伴，茶人之心已
飞向远山远水。

第三章 明代茶道文化整体评价与茶画创作
Chapter Three Overall Evaluation of Tea Culture and Tea−Related Painting Creation of the Ming Dynasty

299

图 192

明·唐寅·《渔溪隐逸图》（局部）·自题诗

　　主人划桨湖中，四野无人，充满恬静气氛，又有某种寂寞而苍凉的味道。虽然画面上只是隐隐约约地看到小渔船上的案几上有两三只瓷杯和红色的紫砂壶，从侧面点画出，茶道修养的孤寂与幽雅。唐寅将钓鱼和烹茶一同作为文人养冶性情的雅事来看待。其同时代的顾璘，"少负才名，……在浙慕孙太初一元不可得见。道衣幅巾，放舟湖上，月下见小舟泊断桥。一僧，一鹤，一童子煮茗，笑曰：'此必太初也。'移舟就之，遂往还无间。"（《明史·文苑二·顾璘传》，中华书局，1974 年，第 7355 页）

图 193

明·唐寅·《琴士图》·台北故宫博物院藏品

　　瀑布、流水、古树、幽琴、煮水、奉茶。琴韵茶香相益相彰。

第三章　明代茶道文化整体评价与茶画创作
Chapter Three Overall Evaluation of Tea Culture and Tea-Related Painting Creation of the Ming Dynasty

301

图 194

明·谢时臣（1487—1567）·《高人雅集图》

　　画高 95.7 厘米，宽 49.9 厘米。

　　1529 年 8 月，王北湄等访作者，所有此作品。蕉竹巨石间，主客三人席地而坐，琴棋书画，茶酒果蔬，文人雅趣，天地和谐。

谢时臣 · 《高人雅集图》（局部）

画中主人坐席上，书画笔墨齐备，茶汤水果置前。茶，为文人所喜好，须臾不可或缺，成为接待宾朋必备之雅物。

图 195

明·许至震·《衡山先生听松图》

嘉靖辛卯岁（1531），为文徵明60岁小像。

图中三种茶壶，分别为手中盘里的、炉子上的、地上茶盘里放的。

图 196

明 · 佚名 · 《西园雅集图》（局部）· 台北故宫博物院藏品

原图高 191.2 厘米，宽 98.3 厘米。

画中内容还是琴棋书画等内容，有专人负责准备茶汤。

图 197

明·佚名·《品茶图》·台北故宫
博物院藏品

　　原画高 157.1 厘米，宽 67 厘米。

　　二人倚树相对而坐，两侍童一汲
水，一生火，准备煎茶。

图 198

明·佚名·《品茶图》·天津博物馆藏品

原图高、宽 30 厘米。宋·李南金云："《茶经》以鱼目涌泉连珠为煮水之节。然近世瀹茶，鲜以鼎镬，用瓶煮水，难以候视，则当以声辨一沸二沸三沸之节。又陆氏之法，以末就茶镬，故以第二沸为合量而下，未若以今汤就茶瓯瀹之，则当用背二涉三之际为合量。乃为声辨之诗云：'砌虫唧唧万蝉催，忽有千车捆载来。听得松风并涧水，急呼缥色绿瓷杯。'其论固已精矣。然瀹茶之法，汤欲嫩而不欲老，盖汤嫩则茶味甘，老则过苦矣。若声如松风涧水而遽瀹之，岂不过于老而苦哉！惟移瓶去火，少待其沸止而瀹之，然后汤适中而茶味甘。此南金之所未讲者也。因补以一诗云：'松风桧雨到来初，急引铜瓶离竹炉。待得声闻俱寂后，一瓯春雪胜醍醐。'"（《鹤林玉露·茶瓶汤候》）

南宋罗大经（1196—1252）所谓"瀹茶"应该就是用沸水冲点之意。其本意是将肉或蔬菜放在沸水锅里煮一下在肉或菜没有熟就很快捞出来，《玉篇》释为"煮也"（《康熙字典》第 610 页）。而明代田艺蘅《煮泉小品》云："芽茶以火作者为次，生晒者为上……生晒茶瀹之瓯中，则旗枪舒畅，清脆鲜明，尤为可爱。"因此宋著作中的"瀹茶"就是点茶，明人瀹茶就是今天我们所谓的冲泡茶。

第三章 明代茶道文化整体评价与茶画创作
Chapter Three Overall Evaluation of Tea Culture and Tea-Related Painting Creation of the Ming Dynasty

307

图 199

明·文嘉·《惠山图》（局部）·上海博物馆藏品

　　文嘉，文徵明之子。此画作于1525年。原画高23.8厘米，宽101厘米。

图 200

明·仇英（约 1498—1552）·《仙山楼阁图》（局部）

中国文人爱茶，但并不自己面面俱到地躬身茶事每一过程。青春时节，桃李灼灼，松柏吐翠。主客在如梦如幻的仙境中，谈天地，说诗歌，和音律，侍者静静地在旁边为他们准备着色香味俱佳的仙人甘露。

第三章　明代茶道文化整体评价与茶画创作
Chapter Three Overall Evaluation of Tea Culture and Tea-Related Painting Creation of the Ming Dynasty

309

图 201

明·仇英·《写经换茶图》（局部）

　　仇英《赵孟頫写经换茶图卷》（局部·图一）画面：图之右前方为赵孟頫在松林树下踞石几写经，似乎才将纸摊开，正待作书。石几前坐有一僧，面向画纸，即是题识上所说的恭上人（中峰明本禅师），而赵孟頫则侧身看着右前方的侍童。此侍童手上捧着一竹茶笼，本书前面第113图《出行图》里，也有一个大小差不多一样大，但无盖笼子。这可能是法门寺茶笼子的延续，相当于《茶经》中的"育"，贮茶用。

图 202

明 · 仇英 · 《人物故事 · 品茗观画》 · 北京故宫博物院藏品

画中最中间端的是青釉瓷茶壶。与真赏斋主人一样，收藏青铜鼎尊为文人所喜爱。

第三章 明代茶道文化整体评价与茶画创作
Chapter Three Overall Evaluation of Tea Culture and Tea-Related Painting Creation of the Ming Dynasty

311

图 203

明·仇英·《东林图》·台北故宫博物院藏品

　　唐寅题跋："抑抑威仪武肃支，乡吾同举学吾师。百年旧宅黄茅厚，四座诸生绛帱垂。灵出尾箕身独禀，器云瑚琏众咸推。佗年抚翼烟霄上，故旧吾当不见遗。年生唐寅。"

图 204

明·仇英·《松溪论画图》·吉林省博物院藏品

　　也可能就是《煮茶论画图》。

图 205

明·仇英·《人物故事图》

平明送客时曙色初露，主人提灯挥手，依惜道别。客船离岸，茶童
已经开始生火备茶。

第三章 明代茶道文化整体评价与茶画创作
Chapter Three Overall Evaluation of Tea Culture and Tea-Related Painting Creation of the Ming Dynasty

313

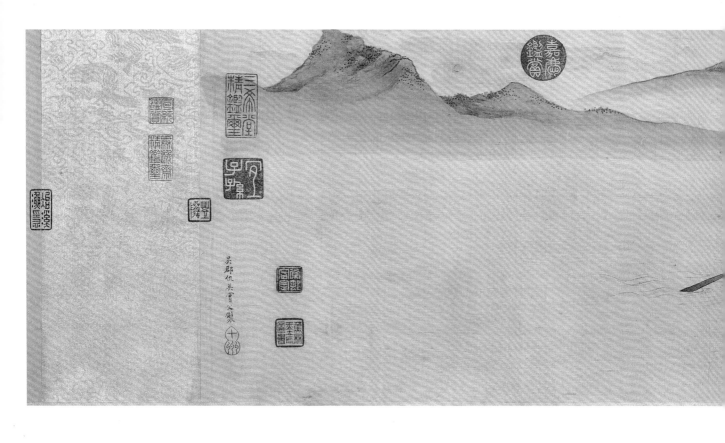

图 206

明 · 仇英 · 《赤壁图》 · 北京故宫博物院藏品

行舟江上，赤壁怀古。烹茶敬英雄，纵论古今事。

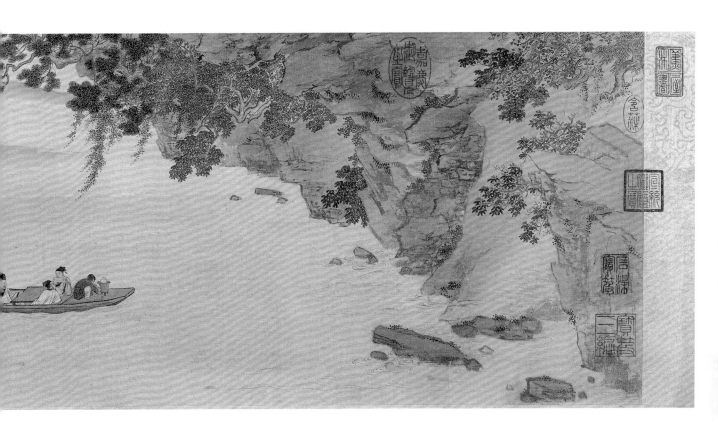

第三章　明代茶道文化整体评价与茶画创作
Chapter Three Overall Evaluation of Tea Culture and Tea-Related Painting Creation of the Ming Dynasty

315

图 207

明·仇英·《临溪水阁图页》·北京故宫博物院藏品

轩临溪潭，品茶钓鱼，修习心性之理相通。

图208

明·仇英·《汉宫春晓》（局部）

　　此为局部，全图都是宫中妇女，表现的是仕女正在解开琴套，准备演奏。黄色木案上的陈设说明过去我们称为温酒碗和酒壶的成套执壶和较大的承碗，应是温茶器。工笔状写宫中嫔妃们的活动：观鸟、浇花、理妆、舞蹈、演乐、观书、对弈、刺绣、熨帛、捕蝶。茶似乎须臾不可缺少。

第三章 明代茶道文化整体评价与茶画创作
Chapter Three Overall Evaluation of Tea Culture and Tea-Related Painting Creation of the Ming Dynasty

317

图 209

明 · 仇英 · 《人物故事 · 吹箫引凤》 · 北京故宫博物院藏品

画王府或皇宫生活，玉女奉茶须拾阶而上。

图 210

明·仇英·《人物故事·高山流水》·北京故宫博物院藏品

主人坐于山房，老仆在房檐口烧水备茶。

图 211

明·仇英·《松亭试泉图轴》·台北故宫博物院藏品

试泉，实是瀹茶来检验水质好坏！

图 212-1

明·仇英·《清明上河图》（局部）之一·西安博物院藏品

　　这是整幅画面最右端，农村人正准备进城。图中上下两个亭子，上面亭子中有穿白衣和穿红衣服的官人叙谈品茶，旁边是一位身穿蓝色衣服的侍者。

第三章　明代茶道文化整体评价与茶画创作
Chapter Three Overall Evaluation of Tea Culture and Tea-Related Painting Creation of the Ming Dynasty

321

图 212-2

明·仇英·《清明上河图》（局部）之二

　　画面中央是一个空间相对独立，距离河边不远的茶肆。在靠近门口的茶舍里，两人对坐一人站立，还有一蓝衣侍者进奉茶盘。三人桌子上是白色茶盏。正在叙话，品茗。这是位于山脚下的茅草屋茶舍。

　　从明晃晃的灯笼看，时间应是黑夜尚未完全褪去的黎明时分。这大约是在"吃早茶"。因为，从各色行人忙碌而急切的情状看，他们是在赶集，而不是离开城市上船回家。三位叙谈，一童子端红色漆果盘奉茶。

图 212-3

明·仇英·《清明上河图》（局部）之三

　　这是刚出城门洞的客船。一伙人正在拉纤，准备泊船。船上二层（画面左侧）有四人在一起，桌上白色茶盏，表明他们在旅途中品茗、叙谈。

第三章 明代茶道文化整体评价与茶画创作
Chapter Three Overall Evaluation of Tea Culture and Tea-Related Painting Creation of the Ming Dynasty

323

图 212-4

明·仇英·《清明上河图》（局部）之四

小船之中，女子品茶。

图 212-5

明·仇英·《清明上河图》（局部）之五

这是位于进入城内的大桥外边的一处茶肆。两桌茶客，围坐木桌，两个穿深褐色衣衫的茶博士正在奉茶、招呼客人。桃红柳绿，杏花灼灼，一派春色盎然。

图 212-6

明·仇英·《清明上河图》（局部）之六

这是入城后第一圆拱桥下面的简易茶肆。四人围坐，白色茶盏，带茶点果子。

图 212-7

明·仇英·《清明上河图》（局部）之七

城门外茶舍。

图 212-8

明·仇英·《清明上河图》（局部）之八

这是东京城中最高档的茶楼！山水树木尽在眼中，来往车马人流均在脚下。

图 212-9

明·仇英·《清明上河图》（局部）之九

　　城市繁华处，顶楼上是士大夫、仕女们聚会品茗之胜境：远眺清清河水、熙熙攘攘各色人群，

细细把玩茶器，用心品尝茶汤。此后或联句抒情，或书画达意。这是当时繁华都市的茶文化。

图 212-10

明·仇英·《清明上河图》（局部）之十

　　典型的茶馆，既有多人围坐的大厅，红色桌椅已经褪色呈褐色；又有两人对桌的小屋，桌椅崭新红亮，临水而居，品茗私语，看鸳鸯戏水，渔夫在河中划船撒网。

第三章　明代茶道文化整体评价与茶画创作
Chapter Three Overall Evaluation of Tea Culture and Tea-Related Painting Creation of the Ming Dynasty

329

图 212-11

明·仇英·《清明上河图》（局部）之十一

　　庭院之外人来人往，屋内室雅人和，茶香袅袅。

图 213

明·仇英·《林溪水阁图》

临溪煮茶是时人雅尚。

图 214

明·仇英·《丽园庭深图》·西安美术学院藏品

　　《丽园庭深图》状写官宦私园内的仕女生活。下部主题为品茗抚琴于角亭；中部以照看儿童戏玩于庭院为主题，两侍女将茶器撤离；上部为主妇观看刺绣于二层楼阁，仕女奉茶。虽然画作是以妇女儿童日常生活为主题，但我们发现，上流社会的妇女生活与品茗息息相关。

图 215

明·仇英·《春游晚归图轴》

　　外出踏春、访友，有琴在怎能没有茶，后一人肩挑都篮！

第三章 明代茶道文化整体评价与茶画创作
Chapter Three Overall Evaluation of Tea Culture and Tea-Related Painting Creation of the Ming Dynasty

333

明·仇英《柳下眠琴图真蹟神品 寒齋藏》

仇英實父製

图216

明·仇英·《柳下眠琴图》·上海博物馆藏品

画高42.595厘米，宽29.097厘米。

高士看稿箋，依琴小憩，待书童拿来画轴、茶灶。琴士不能没有茶喝！

图 217

明·仇英·《蕉荫结夏图轴》

  阮弦、古琴、茶，三人所事不一，但茶
理乐理相合归一。

第三章 明代茶道文化整体评价与茶画创作
Chapter Three Overall Evaluation of Tea Culture and Tea-Related Painting Creation of the Ming Dynasty

335

图 218

明·周翰（1530—1597）·《西园雅集图》（局部）·福建省博物馆藏品

原画高31.5厘米，宽381.5厘米。

我们在明代画里，又看到了唐代就有的碢轴、碾槽与食盒。

图 219

明 · 王耕 · 《坐隐图》（局部）· 北京图书馆藏品

1609 年画，原画高 24 厘米，宽 30 厘米。

三组人分别为备茶、焚香、洗砚。

图 220

明·孙克弘（1532—1611）·《消闲清课图卷·煮茗》·台北故宫博物院藏品

　　从画作名称和题记文句看，他们是究竟佛理，以茶相佐。这也是"禅茶一味"的一
种表现。

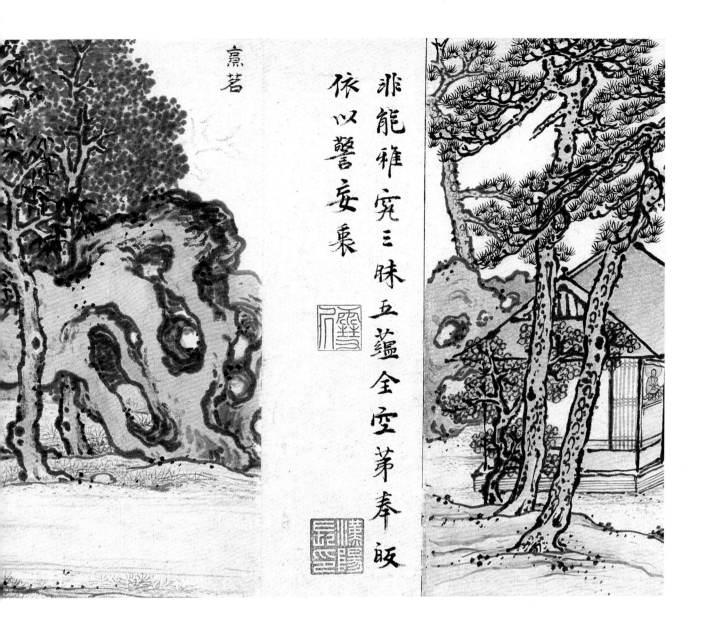

烹茗

非能稚究三昧五蘊全空革奉畈

依以警妄乘

第三章　明代茶道文化整体评价与茶画创作
Chapter Three Overall Evaluation of Tea Culture and Tea-Related Painting Creation of the Ming Dynasty

339

图 221

明·尤求（仇英之后名世）·《园中茗话》·上海博物馆藏品

原画高 32.5 厘米，宽 19 厘米。

梧桐树、珊瑚山、池水，是社会上层得以休闲之处，靠垫、抚几与专用地毯
的使用，似乎告诉人们：上流社会对茶的重视已达到相当程度！

图 222

明·尤求·《坐船烹茗》·上海博物馆藏品

原画高 25 厘米，宽 21.9 厘米。

垂钓、品茗实有异曲同工之妙！中国茶道画，除场面宏大、人数众多的雅集、庭院生活外，多为三好友坐论、两人对语、一人悟对（书写、垂钓）。中国人借茶向内修行，格物致知，最终还要化育众生，资政王道，不是单单渡己的小乘！

第三章 明代茶道文化整体评价与茶画创作
Chapter Three Overall Evaluation of Tea Culture and Tea-Related Painting Creation of the Ming Dynasty

341

图 223

明·尤求·《人物山水》（局部）

　　尤求继仇英名世。人在山水之中，便与山水为侣，茶中亦容纳山水灵气，是人通达
天地的灵物。中国茶道离不开烧水瀹茶。道离不开具象的茶事。茶之道不仅仅把人导向
内敛保守的枯寂，也会引人进入愉悦和合，具有民胞物与的胸怀，了知天人合一至道，
实现天下大同的理想，这是中国茶道的志趣。

图 224

明·尤求·《品古图轴》·北京故宫博物院藏品

　　高 93.1 厘米，宽 36.1 厘米。

　　似欲挥毫，又像在品古物。画面左下为一组完整的茶器。即使茶器只在角隅
点缀，也不可低估茶在文人生活中、社交中的应有地位。（摘自《中国美术史全集》）

第三章　明代茶道文化整体评价与茶画创作
Chapter Three Overall Evaluation of Tea Culture and Tea-Related Painting Creation of the Ming Dynasty

343

图 225

明·丁玉川·《独坐弹琴图》·浙江省博物馆藏品

　　原画高 67.8 厘米，宽 35.7 厘米。

　　焚香、品茗、观书、抚琴，何愁之有？小草小花亦极春，一缕

幽香是清欢。

图 226

明·陆治（1496—1576）·《桐阴高士图》（局部）·天津博物馆藏品

1548 年绘，原画高 93 厘米，宽 49 厘米。

乾隆御题诗："童子烹茶石鼎烟，桐阴日午且高眠。不雕朽木虽鸣鼓，无碍仍称言语贤。乙亥春御题。"左上角自题："茗外寄幽赏，琴中饶濮音。一音洒毛骨，万象开灵襟。蚓窍战水火，月团破璆琳。余音与遗味，优矣澹玄心。"1548 年所作画。嘉靖戊申三月包山陆治制。乾隆二十年（1755）御题诗。

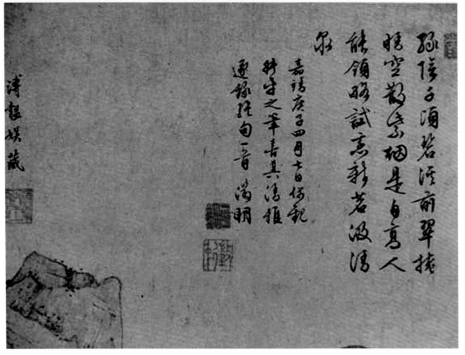

图 227

明·陆治·《竹泉试茗图轴》（局部）

陆治，字叔平，号包山子，师从文徵明。乔木修竹之下，溪流巉岩之旁，两位高士切磋叙语，二童子烧水烹茶。所谓试茗，也是品尝茶味和泉水的品质。画上部有文徵明的题款："绿阴千顷碧溪前，翠掩晴空散紫烟，是日高人能领略，试烹新茗汲清泉。嘉靖庚子四月七日偶观叔平之笔，喜其清雅，遂录绝句一首，徵明。"又有清代摄政王"溥韫娱藏"字样和藏印。

图 228

明·王问（1497—1567）·《煮茶图》

　　王问，无锡人，工人物画，嗜茗。"性不饮酒，而喜啜茗。筑绿萝小径，每遇风清月白，净几明窗，兴至举笔，或书或画，辄写数十幅，如有神助，自谓径丈大字，至老有进。凡仕宦过锡者踵门求见，往往以疾辞。而独好静颐潘君，与之考订书史，唱和篇什，听夕忘倦，其人品高洁可想见已。"（吴智和：《中明茶人集团的性灵生活》引明张萱《西园闻见记》）也许画中表现的就是王问专心烧水瀹茗，潘君正握笔圈点文字，斟酌书法或诗韵仗对。

　　此图系梁友平依据电子版临摹。

第三章 明代茶道文化整体评价与茶画创作
Chapter Three Overall Evaluation of Tea Culture and Tea-Related Painting Creation of the Ming Dynasty

347

图 229

明·周臣·《袍翁咏雪图》（局部）

　　周臣约生活在成化末年至嘉靖元年（1487—1522）间。唐寅、仇英为其学生。白雪皑皑，松柏苍苍，远山茫茫。踏雪来访，万踪杳杳，茗香阵阵。参士会心，天地悠悠。

图 230

明·杜堇（15—16世纪初）画、金琮书·《茶歌》（局部）

卢仝被尊为茶道亚圣，其《茶歌》千古流芳，以其为题材的创作不一而足！
亚圣七碗茶歌当创作于809年的终南山中。

第三章 明代茶道文化整体评价与茶画创作
Chapter Three Overall Evaluation of Tea Culture and Tea-Related Painting Creation of the Ming Dynasty

349

图 231

明·杜堇·《古贤诗意图·听颖师弹琴》（局部）·北京故宫博物院藏品

纸本墨色，高 28 厘米，宽 107.95 厘米。

图 232

明·杜堇·《梅下横琴图》

　　主人在山间阁台抚琴赏梅，书童备茶。

　　陆放翁"高标逸韵君知否？正是层冰积雪时"，把梅花的品格与高士逸君比拟，越是苦寒枯寂，越展示出傲岸不羁的性格。而这与中国茶的精神是一致的：清宁悠远洒脱静雅。高士在寒风中，背靠梅树，横琴赋曲，将茶人自己与梅花融为一体。在茶香幽幽中，尽情抒发"何方可化身千亿，一树梅花一放翁"的古雅情怀。

第三章 明代茶道文化整体评价与茶画创作
Chapter Three Overall Evaluation of Tea Culture and Tea-Related Painting Creation of the Ming Dynasty

351

图 233

明·李士达·《关山风雨图》·台北故宫博物院藏品

李士达，明万历二年（1574）进士。

关山迢遥，更兼风紧雨急。旅人且歇脚，一杯清茶在，消渴提精神。

图 234

明·李士达·《坐听松风图》·台湾故宫博物院藏品

高 167.2 厘米，宽 99.8 厘米。

主人抱腿坐于土地，听风拂松针，水泉潺潺；四童子各司其职：一个扇风煮水，一个侍从准备奉茶，一个采灵芝，一个打开画囊。三足鼎式茶炉列架紫砂壶。红托子配白色茶盏为明代常见搭配。

第三章 明代茶道文化整体评价与茶画创作
Chapter Three Overall Evaluation of Tea Culture and Tea-Related Painting Creation of the Ming Dynasty

353

图 235

明·钱穀（1508—1572）·《品茶
观蕉图》（梁子定名）

图236

明·钱穀·《竹亭对棋图》·辽宁省博物馆藏品

纸本设色，高62.2厘米，宽32.3厘米。

本图题示："小诗拙画问讯凤洲先生。经时不见王青州，养疴高卧林堂幽。竹寒松翠波渺渺，四檐天籁声飕飕。围棋招客赌胜负，劝酬交错挥金瓯。有时弄笔染缃素，句新调古人争收。城居六月如坐甑，曷欲对面销烦忧。美人迢递不可即，东江目断沧波流。丙寅（1566）中秋四日，钱穀。"《竹亭对棋图》以及画上的题诗，可以说是当时苏州文人闲适生活的写照，因而也可以称它是一幅典型的士人消夏图。此图是一幅工笔画，具有很强的艺术感染力，也使得画面的整体感更强。另外，茅亭中画有四个人物，着墨虽不多，但却情态各具。茶炉煮水，以茶驱暑。

图 237

明·钱穀·《惠山煮泉图》(局部)

1570 年，钱穀 63 岁时作品。有
"古稀天子"朱文圆印。望亭，今
苏州西北。

中华茶道图志

Graphical History of Chinese Tea Ceremony

356

图 238

明·钱穀·《竹溪高士图》·西安考古所藏品

　　画中崇山，苍松，翠竹，小溪，清泉；高士们展诗卷，抚古琴，把茶盏；童子，匆匆，烹茶净具。人消瘦，筋骨健，精神爽。不羡黄金罍，不慕暮登台，千古是非，万丈红尘，融化于一瓯碧绿！

第三章　明代茶道文化整体评价与茶画创作
Chapter Three Overall Evaluation of Tea Culture and Tea-Related Painting Creation of the Ming Dynasty

357

图 239

明·钱榖·《定慧禅院图》（局部）·北京故宫博物院藏品

　　1561 年画，原画高 31 厘米，宽 129.3 厘米。

　　《定慧禅院图》对于人们理解茶与佛教关系有借鉴意义。茶与佛教有几个层次的关系。1. 历史关系。茶的发现和使用远远早于佛教的创立。我们认为以东晋杜育《荈赋》等诗文的出现，是中国饮茶文化出现与逐渐形成的标志。但茶从生活化民俗化向仪式化转化过程中，佛教发挥重要作用。唐开元时代（712—741）泰山灵岩寺禅僧以茶入禅并推动茶饮从江南向两京传播发挥积极作用。2. 天下名山僧占多。僧人种茶并发现新茶种，促进制茶方法改进。3. 促进茶道向外传播。最早的是向朝鲜半岛、日本、吐蕃、辽、金、西夏等地传播。京师长安及东都洛阳的国寺、皇寺、名寺及八大祖庭寺院（有的相互重叠）及天台径山寺、五台山等在接待留学僧时连同茶法传授，成为茶道传播的场所和载体。4. 佛教对中国茶道思想和方法的影响。"茶禅一味"的中心意旨是以煮水品茗与习禅开悟相比照。这直接影响到日本，从而有了日本

"佗"的思想。在煮水、投茶、投盐、品尝等一系列动作中，有节奏有讲究，但绝不可有等待有焦虑有期盼。煮水不能想着投茶，投茶时不能想着投盐，而是努力地做到自然的演进。正如得道高僧回答小和尚所问：开悟与不开悟有何不同时，高僧说，虽然都在进行担水、劈柴、烧火，但开悟以后，在担水时，一心一意在担水，而不惦念其他。这是一种训练方式，也是一种修行。因此真正的茶道就是茶人、水与火融和的自然过程。这是茶道的第一个层次。饮茶过程中听音乐、观赏歌舞、对弈、挥毫泼墨、赋诗吟诵等，在既定的茶室焚香、挂画，是茶与其他门类的文艺结合的第二个层次。第三层次，也就是最高的茶道层次，就是将茶道思想与中国传统的儒学价值观及伦理观念和释道审美观相结合，将茶道视为一个综合的文化体系。这个茶道文化体系或称之为中国茶学，同时也是一个需要对历史文化进行发掘，整理同时吸收当代价

值观审美观（包括行为美学）的现代因素，加以整合与重构的新的以茶为载体的文化形态或学术体系。中国自《茶经》《大观茶论》以来的传统茶道著作尚未给出系统的现成的理论体系。在茶事繁盛，但中国茶道思想尚无一致归心情形下，借由禅茶研究，进而建设当代佛教茶文化，成为构建当代中国茶道的可取途径，以释光泉、关剑平、余悦等为代表的茶人对发掘与建设佛教茶文化的努力，志趣高远，步伐坚实。京城谢美霞将吴立民居士发掘的法门寺唐密文化注入饮茶法，独辟蹊径，对当代茶道建设有裨益作用。

图中僧俗对谈，另有书童奉茶。宾主是切磋禅关，还是探讨茶道，我们不得而知，但可以真切地感受到静寂的禅意幽趣。

图 240

明·钱穀·《秦淮冶游图》册页·中国国家博物馆藏品

原图高 23 厘米，宽 40.5 厘米。

品茗、观赏歌舞是士人冶游的重要内容。

图 241

明·陈以诚·《西园雅集图》（局部）·天津博物馆藏品

　　原图高 26.5 厘米，宽 37.96 厘米。

　　侍者偏在一隅，聚精会神地烧火煮水，用耳来判断水温变化。西园雅集是宋以来中国画的传统题材，茶事成为其重要内容。茶往往成为一种媒介，最终归结为思想碰撞、技艺切磋，往往又通过诗、文学表现，而画作的表现则更加形象、生动而直观！

图 242

明·邵徵·《松岳齐年图》（局部）·台北故宫博物院藏品

　　1614 年画，原画高 173.5 厘米，宽 91.2 厘米。

图 243

明·董其昌（1555—1636）·《仿宋元人缩本画及跋》册页

　　董其昌，字玄宰，松江华亭人。举万历十七年（1589 年）进士，改庶吉士。天启五年（1625）拜南京礼部尚书。崇祯九年（1636）卒，年 83 岁。书法师承宋代米芾，画兼宋元诸家之长，自成风格，"潇洒生动，非人力所及也""尺素短札，流布人间，争购宝之。精于品类。收藏家得片语只字以为重。性和易，通禅理，萧闲吐纳，终日无俗语"。（《明史·文苑四·董其昌传》，中华书局，1974 年，第 7396 页）

　　流泉淙淙，高树森茂，屋外蒲团上，友人品茶论道。

图 244

明·戴进（1388—1462）·《春酣图》·台北故宫博物院藏品

　　戴进，钱塘（今浙江杭州）人，字文进，号静庵，又号玉泉山人，善画，幼师叶
澄，及长，得诸家之妙。大率模拟马夏居多。画作展示了春天的踏春游览活动，饮酒，
品茶等场景。

第三章 明代茶道文化整体评价与茶画创作
Chapter Three Overall Evaluation of Tea Culture and Tea—Related Painting Creation of the Ming Dynasty

363

图 245

明·刘俊·《雪夜访普图》·北京故宫博物院藏品

刘俊《雪夜访普图》亦为工笔设色历史人物故事画，款识"锦衣都指挥刘俊写"。内容为宋太祖赵匡胤雪夜拜访赵普，共商统一天下大计的故事。两人雪夜谈话，炉火正旺，白色茶执壶正在加热。其人物、界画，工细严谨，树木枝梢的用笔轻柔，且遒劲潇洒，独具风格。刘俊的生卒年代尚无确切材料可考，根据画风看来可能在宣德前后。此画是他的传世孤本。（穆益勤，明代的宫廷绘画，《文物》，1981年10月7日）

图 246

明·钱贡（16—17 世纪初）·《太平春色》（局部）

　　戊申新正御题诗："村舍康年象，熙熙乐太平。嘉言符我意，名画具真情。炉火围温暖，幼龄侍父兄。应知春色好，墙角早梅荣。"茶执壶在炉火上加热，茶叶可能就在壶内，不断续添新水。墙角梅花与茶，暗香相合，使雪夜平添别样韵味。

图 247

明·项元汴（1525—1590）·《梵林图卷》·南京博物院藏品

　　画高 86 厘米，宽 258 厘米。

　　二层楼阁依山而建，楼下布置佛堂，珊瑚石为景，门外两棵高树，一僧人杖锡缓缓而入，一派宁静恬淡的意境。二楼隐约可见侍童扇火煮水。

第三章 明代茶道文化整体评价与茶画创作
Chapter Three Overall Evaluation of Tea Culture and Tea-Related Painting Creation of the Ming Dynasty

365

图248

明·丁云鹏（1547—1618）·《煮茶图》

丁云鹏，字南羽，号圣华居士，休宁（今属安徽省）人。长于人物、佛菩萨与山水，其绘制墨摸驰名于时。图中人物的穿着和刚刚开放的白中泛绿的玉兰花告诉人们，其时序刚进入草长莺飞的农历三月初，在江南已经是春光明媚的时候，新茶已经上市。主人戴幞头穿白色袍服着薄靴，盘腿坐于树下的床榻之上，在榻外角沿放置用竹筐围护、用石板做承座的炉架，炉上放置黄铜带流的长柄壶，壶底部已经是淡黄色，表明炉火既烧又照，颜色变淡。文士打扮的主人全神贯注地观看长柄壶，侧耳倾听着壶内水温的不断上升，他以三沸判定法把握火候。火候的把握是中国茶道成功的重要一步。主人的前边是低而小的石床。石床上置斗形珊瑚石盆景、茶饼盒、青铜三足鼎、青铜斗形罐、白瓷茶盏置于红茶托上，白瓷高足杯置于红色矮圈足承盘上、黄铜带盖执壶、白瓷罐带青铜色盖。主人前方石床顶端一中年男子正在白色大盆缸内用白瓷碗取水、盆缸的旁边是白瓷带盖瓮。

主人的右方（床榻右端外）以身穿红衣黑布束发的老人双手挽起袖子捧端一敞口、斜腹黑漆木盘，盘内置带木柄的斗笠形笼罩。罩内应该是食物或者茶果。

这幅画的另一参考价值在于：石床顶端带托座的高足杯证明，高足杯就是饮茶器。图中的高足杯口径在十厘米以上。

丁云鹏是中国第一批学习西洋画的国画家，此画色彩艳丽，透视感强，是中国艺术家学习西方绘画的最早成果之一。

图 249

明·丁云鹏·《玉川煮茶图》·北京故宫博物院藏品

　　1612 年画。石案上有高提梁执壶、平顶直盖茶叶罐、泡茶壶。这是作者对唐代茶道亚圣卢仝的想象，将主要表现力集中在玉川子的专注上。

第三章　明代茶道文化整体评价与茶画创作
Chapter Three Overall Evaluation of Tea Culture and Tea-Related Painting Creation of the Ming Dynasty

367

图 250

明·丁云鹏·《滤酒图》

　　虽然画中人物是在过滤酒，但石桌上的器物中有紫砂茶壶、白瓷茶杯。

图 251

明·佚名·《煮茶图》（局部）·北京故宫博物院藏品

　　原轴绢画高 105 厘米，宽 48.5 厘米。

　　从颜色搭配、构图及茶器特征看，即使不是丁云鹏作品，也受到丁云鹏一派的影响。

第三章　明代茶道文化整体评价与茶画创作
Chapter Three Overall Evaluation of Tea Culture and Tea-Related Painting Creation of the Ming Dynasty

369

图 252

明·袁尚统（1570—1661）**·《岁朝图》**

　　此图选自苏州博物馆《明清书画图册》。陆羽提出饮茶须有持久性，"夏兴冬废，非饮也"，饮茶不止于解渴涤烦，而是一种修行。图中人物围桌而坐，品香茶、赏梅韵。

图 253

明·崔子忠（1574—1644）**·《杏园宴集图》·美国私人收藏**

　　崔子忠，字初矑，号石门、石门山人。如果崔子忠是写实叙事，则使人可以联想：唐宋以来的抹茶法，在明末尚有延续。

第三章 明代茶道文化整体评价与茶画创作
Chapter Three Overall Evaluation of Tea Culture and Tea-Related Painting Creation of the Ming Dynasty

371

图 254

明·李流芳（1575—1629）·《临流独坐图》·上海朵云轩藏品

原图高 60 厘米，宽 49 厘米。

作者自题款："墨妙泉清茧纸香，飕飕画秃兔毫芒，挂君素壁心无事，六月梅花凉不凉？夏日兰花大开，芬馥满室，命童子焚香煮茗，涤宋砚，开窗延凉风……时丙寅六月九日，李流芳题。"

图 255

明·士中·《李流芳像》·北京故宫博物院藏品

原画高 124 厘米，宽 40 厘米。

李流芳前边方形石几上有插花、香炉，书童正解去琴护套。角上石茶床上有小铁炉、团扇、白瓷杯、紫砂茶壶（瀹茶），炉上壶专门用以煮水。

第三章 明代茶道文化整体评价与茶画创作
Chapter Three Overall Evaluation of Tea Culture and Tea-Related Painting Creation of the Ming Dynasty

373

图 256

明·蓝瑛（1585—1664）·《煎茶图》（局部）

　　虽然画称《煎茶图》，但主题还是文士友人们的聚会，茶的汤色味道不仅仅是评

论交谈的一部分。

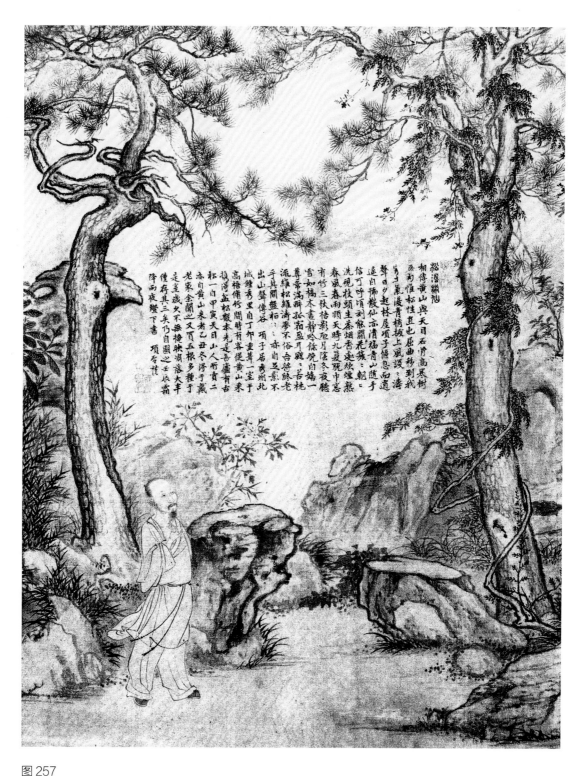

图 257

明·项圣谟（1597—1658）·《谢彬松涛散仙图轴》

　　项圣谟，字孔彰。谢彬，1602 年生，1680 年尚在，卒年不详。项氏植松数棵，自图画，谢彬为其绘画肖像。画中项氏神态自如，风度潇洒，衣带轻飘，右腿欲抬，似在散步。两人绘画珠联璧合相得益彰。项氏自题《松涛散仙》："……朝朝洗砚杖头生，茶烟香处炊烟熟。春风春雨顺天时，九夏脱巾自有竹。三秋梧影照月阴，冬夜听雪如桥木。昼静饮余晚自喜，一尊常满最孤霜……"一人隐居山间，孤寂但不空虚，白天品茗吟诗，夜间自饮杯酒，醉卧孤霜。如果下雪，压得松枝吱吱呀呀，就像桥上的木头不堪重负，嘎吱作响。一杯茶、一杯酒陪伴诗人画家在大自然的怀抱中充满清欣。

第三章 明代茶道文化整体评价与茶画创作
Chapter Three Overall Evaluation of Tea Culture and Tea-Related Painting Creation of the Ming Dynasty

375

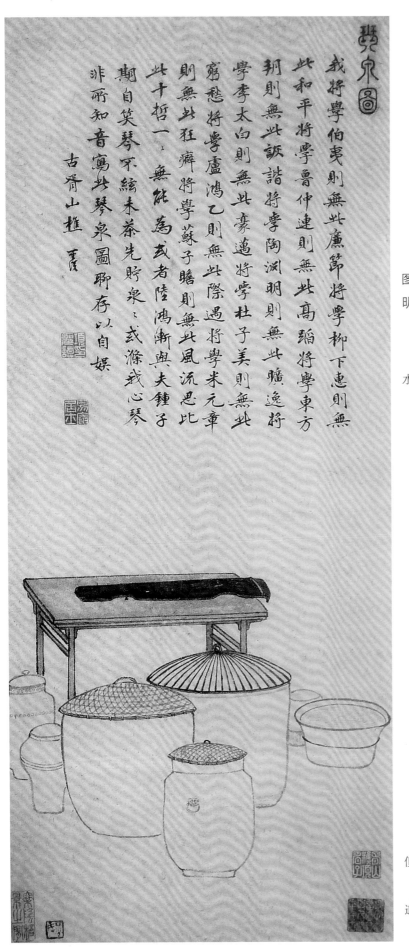

图 258

明·项圣谟·《琴泉图》·北京故宫博物院藏品

画高 65.5 厘米，宽 29.5 厘米。

此图表明，古代爱茶人往往储存泉水以保证品茶水质适合瀹茶。

琴泉图

我将学伯夷，则无此廉节；

将学柳下惠，则无此和平；

将学鲁仲连，则无此高蹈；

将学东方朔，则无此诙谐；

将学陶渊明，则无此旷逸；

将学李太白，则无此豪迈；

将学杜子美，则无此穷愁；

将学卢鸿乙，则无此际遇；

将学米元章，则无此狂癖；

将学苏子瞻，则无此风流。

思比此十哲，一一无能为；

或者陆鸿渐，与夫钟子期；

自笑琴不弦，未茶先贮泉；

泉或涤我心，琴非所知音；

写此琴泉图，聊存以自娱。

古胥山樵　圣谟

钤印："项圣谟印"等。项氏贫困，卖画为生，但意有不可，即使玉帛羊雁，也不出售。

实际上，我们可以将此自题视为明代士大夫对茶道思想的诠释：既不过于旷达，更不狂癖，而要达到"中庸、清和、朴素、清雅"的至高境界。

图 259

明·陈洪绶（1599—1652）·《品茶图》

　　陈洪绶，杰出的画家，人物画尤为突出。他在紫砂陶艺方面的贡献也为后人称道。《品茶图》告诉我们：这时的茶道模式属散茶瀹茶道。所用烧水器是铜质壶，粗短直流，半圆形执手。这类壶造型在考古发掘中没有形制相同或相似的瓷壶出土。因此我们判断当是铜质壶或铅锡壶，但铅锡不耐烧，很可能就是铜质壶。自题款："老莲洪绶画于青藤书屋"。

第三章　明代茶道文化整体评价与茶画创作
Chapter Three Overall Evaluation of Tea Culture and Tea-Related Painting Creation of the Ming Dynasty

377

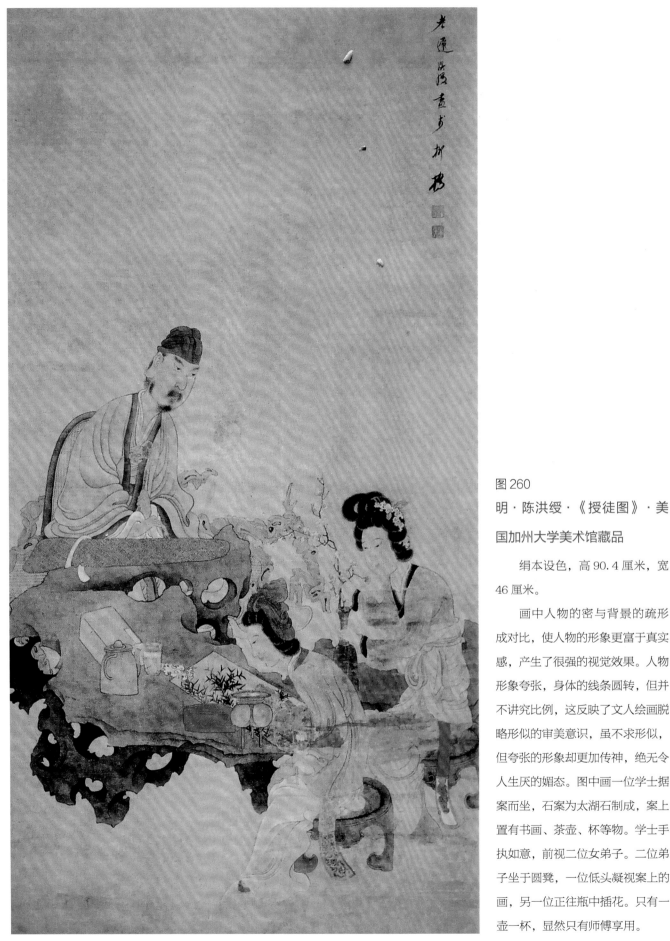

图 260

明 · 陈洪绶 · 《授徒图》 · 美国加州大学美术馆藏品

绢本设色，高 90.4 厘米，宽 46 厘米。

画中人物的密与背景的疏形成对比，使人物的形象更富于真实感，产生了很强的视觉效果。人物形象夸张，身体的线条圆转，但并不讲究比例，这反映了文人绘画脱略形似的审美意识，虽不求形似，但夸张的形象却更加传神，绝无令人生厌的媚态。图中画一位学士据案而坐，石案为太湖石制成，案上置有书画、茶壶、杯等物。学士手执如意，前视二位女弟子。二位弟子坐于圆凳，一位低头凝视案上的画，另一位正往瓶中插花。只有一壶一杯，显然只有师傅享用。

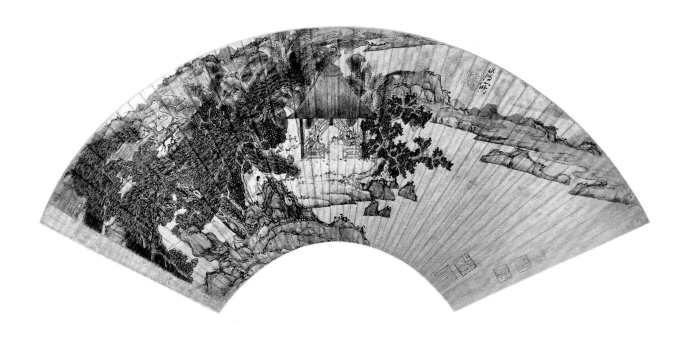

图 261

明·陈洪绶·《烹茶图》

　　扇面画。尺余见方，但山水亭阁、湖水布局典雅，人物动

感十足。

第三章　明代茶道文化整体评价与茶画创作
Chapter Three Overall Evaluation of Tea Culture and Tea-Related Painting Creation of the Ming Dynasty

379

图 262

明·陈洪绶·《玉川子像》·程十发藏画陈列馆藏品

画中人物表情凝重，王川子（卢仝）喜看书、善思考、嗜茶的特征得到很好的表现。在长安作画，是对"亚圣"的纪念。

图 263

明·陈洪绶·《谱泉》（局部）·台北故宫博物院藏品

原画高 21.4 厘米，宽 29.8 厘米。

铜壶煮水、紫砂瀹茶。一人独品为神。"谱泉"之意，是对宜茶之泉排出谱系。

图 264

明·陈洪绶·《侍女人物图》·美国
伯克利加州大学美术馆藏品

原画高 90.4 厘米，宽 46 厘米。

作者自题："道心，韵事，平生自
许；名花美人晨夕与处，为仙耶！人耶？
吾目中少见此侣 洪绶题。"我们看到，
陈洪绶与陆羽有相似点：非僧非俗，即
僧即俗。因时代不同，对于名花美人大
加赞赏，能晨夕相处引以为豪。题记中
"道心韵事"正反映了作者视追求艺术、
感悟佛理道经为道心，赏花品茗为雅事。
不论是在形而下的日常生活中，还是在
形而上的艺术追求中，茶已深深地融入
晚明士人的生命之中。

图 265

明·陈洪绶·《闲话官事图》·沈阳故宫博物院藏品

原画高 92.4 厘米，宽 46.8 厘米。

第三章 明代茶道文化整体评价与茶画创作
Chapter Three Overall Evaluation of Tea Culture and Tea-Related Painting Creation of the Ming Dynasty

383

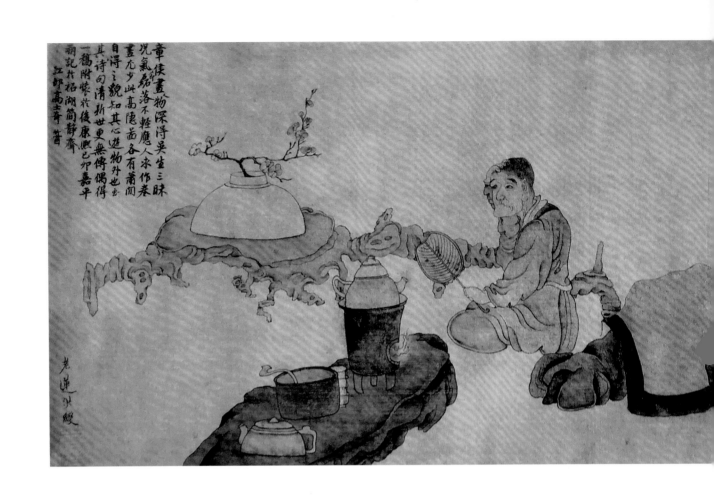

图 266

明 · 陈洪绶 · 《高隐图》（局部）· 王己千藏品

　　约 1647 年画，原画高 30 厘米，宽 142 厘米。

图 267

明·陈梦鹤·《刘宗周像》·上海博物馆藏品

原画高 131.3 厘米，宽 57 厘米。

刘宗周（1578—1645），绍兴府人。

第三章 明代茶道文化整体评价与茶画创作
Chapter Three Overall Evaluation of Tea Culture and Tea-Related Painting Creation of the Ming Dynasty

385

图 268-1

明 · 佚名 · 《入跸图》（局部）

　　画高 92.1 厘米，宽 3003.6 厘米。

　　《汉书》云：出称警，陆路行；入谓跸，水路走。图中主人有明世宗和明武宗两种说法。专为准备御馔的小船紧随天子龙舟之后，茶酒执壶、茶盏酒杯各有不同，时花仙桃陈列有序。

图 268-2

明·佚名·《入跸图》（局部）

　　画之局部是下船以后陆地行走情形：两人抬着苫着黄丝绸的座椅，另一人抱着黄釉密
封盖口的酒瓶茶罐，另一人提着黄丝绸包裹的套盒，大约为茶器具。

第三章　明代茶道文化整体评价与茶画创作
Chapter Three Overall Evaluation of Tea Culture and Tea-Related Painting Creation of the Ming Dynasty

387

图 269

明·佚名·《南都繁会图卷》·南京

　　在南京城内，简易茶馆到处可见，人们品茶听曲，安逸自在。

图 270

明·佚名·《皇都图卷》·北京

　　"（钟）惺字伯敬，竟陵人，万历三十八年（1610）进士。……官南都，僦秦淮水阁读史，恒至丙夜，有所见即笔之，名曰《史怀》。晚逃于禅以卒。"（《明史·文苑四·钟惺传》，中华书局，1974，第 7399 页）都市之中，茶肆林立，付钱取饮，极为方便。

图 271

明·佚名·《西园雅集图》·西安美术学院藏品

原画高 825 厘米，宽 1015 厘米。

《西园雅集图》在中国画史上多次出现。是对历史典故和历史人物的追忆缅怀，表达文人雅士对高雅的诗书画及茶道生活的向往和推崇。西园雅集，是北宋英宗赵曙的驸马、山水画大家王诜在北宋元祐二年（1087）在汴梁府邸——西园，邀苏轼、苏澈、黄庭坚、米芾、蔡肇、李公麟、李之仪、秦观、晁补之、刘泾、张耒、王钦臣、陈景元、郑嘉会、圆通大师等主、友十六位，在"宝绘堂"赏珍观奇之后，在园内赋诗唱和、品茗论道的情形。画中六组人物各自活动，分别可名为布道、石壁题字、弹弦听音、即兴诗文创作、绘画、咏诵听讲诗文。左上角为圆通大师焚香布道。主人应是占居画面中心坐于主桌右端者，夫人侍女，观看友人高执管悬腕书写；另一红木条桌上，一位友人正在颜料盏中润笔。还有一组是弹奏欣赏音乐，最后一组是一人坐于蒲团，一人斜石头靠，咏诵聆听诗文。在这六组十八人（含二夫人侍女）之外，还有四位服务的书童。在圆通法师正下面是两个侍童，分别用勺舀水、用瓶运水，画面中不上端人物为侍童吹火煮水。另一个书童，身穿蓝衣，左手抱卷轴，右手提高提梁茶壶。无疑这四个侍童是为主客冲泡茶水的。品茶已经深入地融进了文人士大夫文化创作与文化活动之中。

后代画家往往以其生活时代的形式表现前代或历史人物的史实，他们重视气氛营造和意境的把握，而对日常生活用度方面各种器物及其使用方式的变化却往往忽略。甚至画家笔下的人物服饰都不是画中人物那个时代的，可能就是画家生活的时代的服饰及其搭配方式。

第三章 明代茶道文化整体评价与茶画创作
Chapter Three Overall Evaluation of Tea Culture and Tea-Related Painting Creation of the Ming Dynasty

389

图 272

明 · 余令 · 《煮茶闻道图》 · 西安美术学院藏品

原画高 30.2 厘米, 宽 20.5 厘米。

款题: "定州郎余令绘", 右下角钤 "顾氏秘藏书画" (白文长方印)、 "项子京家珍藏" (朱文长方印)。画面由中年师傅、书童和煮水炉组成正三角构图。画面主体位置为师傅盘腿坐于兽皮坐垫, 左手搭左膝盖, 右手轻捻稀疏的胡须, 正在向书童叙说茶汤成败得失。书童蹲坐于地上, 双手端紫砂壶。从师傅身后紫色茶几上花瓶内的玉兰花看, 正是三月新茶摘采季节。三足风炉上的高直提梁紫砂壶带盖, 带流, 告诉人们煮水紫砂和瀹茶紫砂有不同的用途和分工。画面给人一种怡然宁静、平和恬淡的气息, 而这正是茶道所营造的精神氛围。作者生卒不详, 生平行状也不详。

图273

明·吴彬（1573—1643）·《文士雅集图》·西安美术学院藏品

　　于1597年所绘，高110.8厘米，宽45.9厘米。

　　也是对北宋元祐二年（1087）驸马王诜邀约苏轼、苏澈及米芾、秦观等十余人与汴梁西园雅集的回顾与憧憬。主客分六组活动，主席为王诜陪友人观看书画、听讲诗文、挥笔作书画；围棋对弈；僧道探讨音律；投筹竞赛；赏花；观石。图中人物除二十二位参加活动的外，包括侍者、书童、夫人及侍女，共三十六人之多。

　　此图状写深深庭院中丰富而情趣的妇女与儿童的生活。宽阔严整的围墙将外界隔开形成独立完整的一个区域，构成不闻鸡鸣狗吠，不见商贾小贩的独立空间。这与芸芸大众形成对照：华贵、恬淡、富足而光鲜。画面有二层轩窗式阁楼、阁楼一楼及院落与四角亭子三部分组成。女主人在二楼观看欣赏织女刺绣，婢女奉茶于后。院中主体为小儿嬉闹，有提茶壶端茶盘的婢女走过。在四角亭中活动的是一夫人抚琴，一妇人认真聆听，二婢女在后。琴前放一个直筒高圈足白瓷茶杯。我们看到，饮茶，已经是上流社会的浓厚风尚和日常生活的一部分。

图 274-1

明·黄卷·《嬉春图》（局部）·上海博物馆藏品

　　1636 年画，原画高 38 厘米，宽 311.2 厘米。

　　泛舟湖上，品茗叙话，是中国茶道的理想图景。

图 274-2

明·黄卷·《嬉春图》（局部）·上海博物馆藏品

五人各演奏乐器，三人专门准备茶水。

满族祖先可追溯到五代时女真人，他们建立的金与宋和辽有紧密关系，茶曾是其间重要纽带。十七世纪中叶建立的清朝在文化上仍然推崇儒学为核心的汉文化，也深受宋明茶文化影响，并且继续推行『榷茶易马』政策。茶事进入新的繁荣昌盛时期，茶成为清朝对外交往的重要媒介。茶叶种植面积扩大、新的茶品纷纷涌现，贡茶种类增多，地方茶文化方兴未艾，积极应对工业革命带来的国际贸易版图的改变成为清朝茶界的时代特征。中国茶叶以空前的规模和速度走向世界，但茶道文化却并没有随之扩大传播，这是应引起反思的茶史内容。

第四章

清代茶文化的整体
风貌

Chapter Four
General Feature of Tea Culture
in the Qing Dynasty

清朝由先前的女真金朝发展而来，其茶文化必然带有唐宋余韵蒙元气息。像前朝一样，宫廷是社会风尚的最高展示场所，同时也影响着社会风尚的变化与发展。沈阳、北京和台北三大故宫保留着清朝的档案和文物，这些是我们研究清代茶文化的最直接材料。

沈阳故宫博物院栾晔《从沈阳故宫藏清代宫廷茶具看清宫饮茶风尚》："清制，朝廷每岁新正均要于宫内举行大型的茶宴，如在康熙后期至乾隆年间的鼎盛时期，宫中曾多次按例举行筵宴，这其中均有茶饮名列其中。""根据清宫前后所设置的各个与茶品有关的部门，可将宫内茶房分为两类：一类为收贮茶品的茶叶库房，以固定藏贮各类进贡的御茶；另一类为制作茶饮的组织，以专门烹茶煮茗，供给帝后、皇室成员和其他入宫品茶者饮用。"[1]

"清宫饮茶用具品种众多，制作精美，造型高雅，质料丰富……有茶壶、盖碗、茶杯、茶盅、茶盏、茶盘、茶船、茶罐等，各类茶具中尤以壶器类型为最多，

Qing Dynasty took its form on the basis of Kin Dynasty established by Nuzhen people, so the tea culture of the Qing Dynasty must have had the elements and influence from the Tang, Song, Yuan dynasties. The royal court of the Qing Dynasty, like the former dynasties, was a show place of the highest level for social fashion, at the same time also affected the change and development of the social fashion. The archives and cultural relics of the Qing Dynasty preserved in the three palace museums in Shenyang, Beijing, and Taibei are the first-hand materials for our study on tea culture of the Qing Dynasty.

Luan Ye from Shenyang Palace Museum said in his article "Tea Drinking Fashion of Qing-Dynasty Royal Court Reflected by Palace Tea Wares Preserved in Shenyang Palace Museum": "A large tea feast was held in imperial court palace in the first month of lunar new year every year according to the Qing Dynasty regulation. For instance royal banquets with names of tea on its menu had been held for many times in the heyday of the late Kangxi Reign to the Qianlong Reign.""The imperial palace tea room could be divided according to the tea-related departments set up in the Qing Dynasty palace before and after into two categories: tea warehouse for storage of tributary imperial tea and organization for specifically cooking or boiling tea for emperor, royal household and other tea tasters in the imperial palace."[1]

Luan Ye's article shows: "The variety of tea wares with elegant shape, exquisite quality and rich material used in the Qing Dynasty royal court was numerous such as teapot, covered bowl, tea cup, tea calyx, tea tray, ship-shaped tea vessel, tea jar, etc. Pot type took up most of all kinds of tea ware including flat shape pot, high base and straight handle

有扁平形、高桩端把式、提梁式、竹节式、石榴式、佛手式、桃形倒流式、人形、鸟形及做成福、禄、寿、喜字形的，还有一些独具特色的菊瓣式、瓜式、梨式、莲子、方体和直流式等茶壶。"作者认为"清宫茶风延续了中国茶文化的发展，使传统文化由清代而顺利递进到现代，这应是其对中国古代文化和茶文化的最大贡献"。[2]

作者认为"清宫茶风延续了中国茶文化的发展，使传统文化由清代而顺利递进到现代，这应是其对中国古代文化和茶文化的最大贡献"。

抚顺师范高等专科学校人文系马东在《从〈清稗类钞〉看清代茶文化发展的广度和深度》一文中认为，"《清稗类钞》中记述的饮茶阶层，除文人外，上至皇室大臣，下至平民百姓，涵盖清代社会生活中的各色人群，从中可清晰地看到清代茶饮文化在阶层分布上具有广泛性"，"乾隆、光绪、孝庄后、孝钦后等均有嗜茶喜好"[3]。"清代茶馆上承晚明，在数量、种类、功能上皆蔚为大

pot, loop handle pot, bamboo-shaped pot, pomegranate-shaped pot, Buddha's-hand-shaped pot, peach-shaped backward flowing pot, human-shaped pot, bird-shaped pot, Chinese-word-shaped pot, some unique chrysanthemum petal type pot, melon type pot, pear type pot, lotus seeds type pot, cube type pot, straight-flowing pot, etc."[2]

The author thinks that "Tea drinking fashion in the Qing Dynasty royal court continued the development of Chinese tea culture, passed it smoothly from the Qing Dynasty to modern times. This should be the biggest contribution the Qing Dynasty made to China's ancient culture and tea culture."

Ma Dong from Social Study Department of Fushun Normal College wrote in his article: "See the Breadth and Depth of Development of Tea Culture in Qing Dynasty from *A Collection of Stories and Hearsay of Qing Dynasty*" that "The tea drinking classes recorded in *A Collection of Stories and Hearsay of Qing Dynasty* include people from all walks of life in the society such as literati, royal household, ministers, and plain citizens.""The Emperor Qianlong, Emperor Guangxu, Queen Xiaozhuang, and Queen Xiaoqin were addicted to tea."[3] "The tea-house saw its period of great prosperity in the quantity, type and function in the Qing Dynasty on the basis of the late Ming Dynasty and played an indispensable role in daily life of all people."[4] But it's a pity that the expression "tea culture" did not appear once in the full text.

A Dream in Red Mansions is on the top of the list of four great Chinese classical novels. Its author Cao Xueqin was not only proficient in poetry, calligraphy, and painting, but also paid most careful attention to tea culture and understood the core connotation of tea culture in his unique way. The

观，完全融入了中国各阶层人民的生活。因此，可以说茶馆的真正鼎盛是在清朝。"[4] 可惜全文洋洋洒洒，似乎没有见到"茶道"二字出现。

《红楼梦》是古典四大名著之首，作者曹雪芹不仅精通诗书画，对茶文化特别留心，对茶道三昧也颇有体悟。《红楼梦》成为反映茶道文化最深刻的清代著作。虽然是小说，没有对茶道文化进行系统性的解说与论证，但却便于从品茗的宏观环境铺排铺垫和场面的描写，从细致入微的细节展示茶作为人情文脉的特殊媒介作用。

清代茶学代表性著作是陆廷灿的《续茶经》，"把历代茶文献按《茶经》之体例，分门别类，摘分编录，类辑成书，从而把唐代以来茶文化的发展历程梳理得极为清楚，并全景式的展示在读者面前。这就表明，《续茶经》不仅是名副其实的《茶经》续篇，而且堪称中国古代的茶叶百科全书"。[5]

作者官至福建武夷山所在的崇安知县。暇时习茶事，于采摘、

novel reflected most profoundly Chinese tea culture not by the means of systematic explanation and demonstration, but detailed description on macro tea-drinking environment, the special function of tea as a medium of human feelings, and related cultural context in the Qing Dynasty.

Supplement of Tea Classic by Lu Tingcan was a representative work on tea science of the Qing Dynasty can be regarded as an ancient encyclopedia on Chinese tea because it classifies tea-related documents and literary works from the Tang Dynasty to the Qing Dynasty according to the stylistic rules and layout of *The Classic of Tea*. It shows readers a panoramic view of the development of tea culture since the Tang Dynasty."[5]

Lu Tingcan, the author of *Supplement of Tea Classic*, was the magistrate of Chong'an County at Mount Wuyi in Fujian Province doing tea-related things in his leisure time such as picking tea leaf, steaming and baking tea leaf, tasting making-tea water, controlling the time of boiling water on fire, etc. He managed to find the origin of tea history and answers for tea-concerning questions by asking others or looking up books. So his work on tea was not simply the compilation of data or lyrical literary fiction of tea, but the crystallization of personal practice and careful research. Although the following books were not included into his work *Tea Manual* by Zhu Quan, *Tea Manual* by Zhu Youbin, *On Tea-horse Barter Trade* by Chen Jiang, *Supplement of Tea Manual* by Zhao Zhilv, etc. Although the record of tea paintings of the past dynasties in the "Chart as Tenth Section" of *The Classic of Tea* by Lu Yu is not comprehensive enough, this book still represented the highest level of research on the cultural history of ancient Chinese tea ceremony.

Zhen Jun praised highly Biluochun tea from Suzhou, Jie

蒸焙、试汤、候火之法，益得其精。究悉源流，每以茶事下询，查阅诸书。因此其著作并非是馆阁资料汇编，亦非望茶抒情的文学杜撰，是实践与研究的结晶。虽然漏记朱权《茶谱》、朱祐槟《茶谱》、陈讲《茶马志》、赵之履《茶谱续编》等著录；《茶之图》关于历代茶画的记载，也不够全面。但他代表了当时茶道文化史的最高研究水平。

震钧《茶说》推崇苏州碧螺春、芥茶和龙井茶。云："四煎法。东坡诗云：'蟹眼已过鱼眼生，飕飕欲作松风鸣。'此言真得煎茶妙诀。大抵煎茶旨要，全在候汤。酌水于铫，炙炭于炉，惟恃韛韛之力，此时挥扇不可少停。后细末徐起，是为蟹眼；少顷，巨沫跳珠，是为鱼眼；是则微响初鸣，则松风鸣也。……自蟹眼是及出水一二匙，至松风鸣时复入之，以止其汤，即下茶叶。大约铫水半升，受叶二钱。少顷水再沸，如奔涛溅沫，而茶成矣。然此际最难候，太过则老，老则茶香已去，而水已重浊；不及则嫩，嫩则茶香未发，水尚薄弱，二者皆为失

tea, and Longjing tea in his work "Tea Talk". "There are four methods of judging the boiling water for making tea. Su Shi mentioned them in his poem:"The early-stage boiling water bubble looks like crab eye, the second-stage boiling water looks like fish eye sounding like the soughing of the wind in the pines.' I believe that Su Shi understood the secret of making tea which should be controlling the time of boiling water. Put the water in a pan and charcoal in stove, do not stop waving your fan when steam rise; then very tinny bubbles look like crab eye gradually appear, after a short while come big bubbles looking like fish eye; at this time the sound is heard like the soughing of the wind in the pines... Lade one or two spoons of water out of the pan when there are tinny bubbles, pour the lading-out water back into the pan when there are big bubbles in order to cease the bubbles and with that place the tea leaf in the boiling water, then a small amount of tea leaf into the boiling water while the water bubble comes again. The tea is done when the water boil again with big bubbles. However this moment is very hard to know, the fragrant smell of the tea goes away and the water becomes turbid if you find it too late, the fragrance of the tea has not come out fully and the water is a bit tender thus weak if you find it too early."[6] He also said, the tender water gave throat a feeling of light and hollowness while tough water a feeling of weight hence difficult to swallow.

"Tea tastes best made with three-times boiling water giving our throat a feeling of composure and clearness at the same time giving our tongue a feeling of emptiness as nothing in there. You had mastered the key skill of making tea if you would have succeeded in doing this."[7] As to the atmosphere of environment for drinking tea, Zhen Jun said: "It should be at sweet hours of a wonderful day in one's spare time

饪。"[6]他还说，水嫩则入口觉其质轻而不实，水老下喉觉其质重而难咽。

"惟三沸水已过，水味正妙，入口而沉着，下咽而轻扬，挢舌试之，空如无物。火候至此，至矣！"[7]对于品茗的环境氛围应是"良辰胜日，二三知己，心闲手暇，清淡未厌，出而效技"。[8]震钧此文短短三千六百言，但我觉得，他却继承了唐宋茶道的核心理念。并将烹茶视为一种技艺，二三知己一起是"效技"。对水的品鉴和味道的把握，可谓丝毫有别，锱铢必较，达到艺术的境界。

余怀（1617—？），福建莆田人，康熙戊午著《茶史补》，记茶事六十三事，其中《沙苑侯传》仿苏轼《叶嘉传》写紫砂壶："吴越之间，高人韵士，山僧野老，莫不愿交于侯。侯亦坦中空洞，不择贵贱亲疏，倾心结交，百余岁以寿终。"赞美茶叶"涤烦荡秽，清心助德"。

除了茶事著作外，文人诗赋也成为我们考察一时代茶道文化的重要参考。

我们从以乾隆为代表的皇室

inviting two or three close friends and making tea with joy and satisfaction for them with one's skills."[8] Zhen Jun's article contains just 3600 words, but I think he inherited the core concept of tea culture of the Tang and Song dynasties. He considered the tea-cooking as a skill and found the slightest differences in the tasting of water and tea which attained the artistic realm.

Yu Huai (1617-?), was born in Putian, Fujian Province, wrote *Supplement on Tea History* in 55th year of the Kangxi Reign recording sixty three tea-related events among which "Biography of Shayuan Marquis" section was written by imitating "Biography of Ye Jia" by Su Shi about purple clay teapot. "In the Wu-Yue area (today's Jiangsu and Zhejiang area), talented persons, refined scholars, Buddhist monks, and old hermits were all willing to get along with Shayuan marquis. Shayuan marquis also treated all his friends equally with wholeheartedness, frankness, and open mind. He died at age over one hundred." Yu Huai praised the tea "Can chase away annoyance, wash away dirt, clear one's mind, and help cultivating virtue."

In addition to the tea-related books, poetry and prose by scholars are important references for our study on tea culture of an era.

Observation on royal family and aristocrats represented by emperor Qianlong and on scholars represented by Qian Qianyi and Yuan Mei shows that if Emperor Qianlong was a real political leader, then Emperor Huizong of the Song Dynasty was definitely an emperor of artistic realm. But there was one thing in common between them which was tea meaning they were both tea lover. Emperor Huizong valued highly of Beiyuan tea from Fujian Province. Encouraged by treacherous court official Ding Wei, and gentleman

贵族和以钱谦益、袁枚为代表的文人来观察。如果说乾隆皇帝是真正的政治领袖，那么宋徽宗是艺术王国里的皇帝。但有一点他们是相同的，这就是茶，他们都钟情于茶。宋徽宗特别推崇福建北苑茶。在佞臣丁谓和君子蔡襄的先后鼓动下，他特别喜欢龙团凤饼，对斗茶也特别在行。与宋代茶道审美相适应，建窑黑色茶盏是斗茶的高级茶器。追求茶沫雪白色，因而黑色盏反差最大。日本人最早得此类盏是通过浙江天目山寺院。也把建盏称为天目器、天目盏。有兔毫盏、滴油盏和曜变盏三类。但从馆藏和发掘品看，也有纯黑色的。

乾隆皇帝与徽宗中间还隔了元明两代。这就自然处在中国茶道发展的又一个阶段——炒青瀹茶。他使用和喜好的茶器以瓷器和紫砂壶为主。在北京香山、万寿山、颐和园及河北承德避暑山庄等地建有十多处茶舍——专门喝茶的亭、阁、馆、厦，例如香山"玉乳泉"、清漪园"春风啜茗台"、圆明园"清辉阁"、万寿山"清可轩"、避暑山庄"千尺雪"和"味甘书屋"，

Cai Xiang. Emperor Huizong especially liked "Coiled-dragon-and-phoenix Pie" tea and also was very good at tea competition. Black glazed porcelain tea calyx from Jianzhou kiln in Fujian Province was the high grade tea ware which formed a maximum contrast with snow-white tea foam in tea competition of the Song Dynasty which was adjusted with aesthetic criterion of tea culture. Jianzhou kiln calyx was first introduced into Japan through Tianmushan Buddhist Temple in Zhejiang Province. Japanese gave it another name "Tianmu Ware" or "Tianmu Calyx". Jianzhou kiln calyx included three types— Hare's Fur Temmoku calyx, Oil-drop calyx, and Yohen Tenmoku calyx. Pure black glazed wares were also found in the museum collection and excavation.

There were two dynasties separated from emperor Qianlong and emperor Huizong. So, Qianlong Reign of the Qing Dynasty naturally fell into a period of drinking stewed tea with pan-fire-processed tea leaf in the development of Chinese tea culture. Emperor Qianlong loved and mainly used porcelain and purple clay tea wares, gave order to build more than ten tea houses with booths, pavilions, guesthouses, mansions at Xiangshan Hill, Wanshou Hill, the Summer Palace in Beijing, and Chengde Summer Resort in Hebei Province.For instance, Yuruquan Pavilion at Xiangshan Hill, Tea-drinking Platform and Qinghui Pavilion at the Summer Palace, Refreshing Room at Wanshou Hill, Snow-white Waterfall and Sweet Flavor Reading Room at Chengde Summer Resort, and Heart-east Room at Beihai Lake.

Besides, Qianlong was the emperor who created the largest amount of tea poems in Chinese history.

Yixing purple clay tea wares, tea spoons and tea bowls were made by emperor Qianlong's order at Anning Branch of Suzhou Weaving Bureau containing Yixing purple clay

还有西苑焙茶坞（北海静心斋）。

此外，乾隆皇帝是创作茶诗歌最多的皇帝。

他还御制宜兴紫砂茶器，御制茶诗茶碗。其茶器，多由苏州织造安宁所承办。有竹木（漆器、檀香）茶籝、茶床、宜兴壶、茶叶罐、茶盅、木盏、茶盘、香几。

在重要节日举行茶会，并赐赏大臣。

珍藏文徵明《事茶图》等茶道名画，和韵历代著名茶诗。可以说乾隆朝将中国茶道文化推向一个新高峰。室、备、温、置、赏、沸、吟、闻、浸、敬、赐、品十二个程序，是乾隆宫廷茶的既定模式。但可随时事不同，会有所增减。他喜欢龙井、日铸，也自创三清茶。

可以看出，以乾隆皇帝为旗帜的清代茶道文化具有包容和创新精神。和雅、亲和、悟真、礼敬、精进的精神融合在清代茶事之中。

由于帝王的积极倡导，清代的茶事活动非常活跃，显示出历史新高度：茶器精益求精、当代社会具有的各类名茶在清代已经大体成型在绿茶基础上乌龙茶和紧压茶受到人们的喜爱。

tea pot, also tea box, tea couch, tea jar, handless tea cup of bamboo or wooden material, wooden calyx, tea tray, and long table as incense burner stand.

These tea wares were utensils for tea party at important festivals also presents given by the emperor to his ministers.

Emperor Qianlong had in his personal collection famous paintings including one by Wen Zhengming called "Tea Activity Painting" with the same rhythm of renowned tea poems of past dynasties. So to speak, the Qianlong Reign of the Qing Dynasty pushed Chinese tea culture to a new peak. The royal court tea was usually prepared according to fixed mode covering twelve procedures: selecting place, preparatory work, warming tea wares, placing tea wares, appreciating, boiling water, chanting, smelling, dipping, worshiping, granting, and tasting. But the procedures sometimes were decreased due to the practical condition. Emperor Qianlong was fond of Longjing tea and Rizhu tea which are both green tea, even invented a kind of tea drink with Chinese herbal medicinal ingredients.

It was obvious that tea culture of the Qing Dynasty with emperor Qianlong as its representative had the spirit of inclusiveness and innovation, also harmony, elegance, affiliation, truth, respect, enhancement which had been fused in the tea activities.

Because of emperors' active initiating and guiding, the Qing Dynasty saw frequent and lively tea activities at a new historical height. Tea ware were made in principle of perfection.All kinds of famous tea of contemporary society had taken its shape in the Qing Dynasty on the whole, oolong tea and compressed tea besides of green tea were loved by people of the time.

Qian Qianyi (1582-1664) wrote "Verse on Gratitude for Receiving High Quality Tea from Friend": "Things change

钱谦益（1582—1664）"世事
突兀看枪旗，富贵纷纭诧团饼"（《谢
于昭远寄庙后茶次东坡和钱安道
韵》），"高山流水在何许？但见
风轻花落蒸茶烟"（《戏题徐元叹
所藏钟伯敬茶讯诗卷》）表现了以
茶寄情充满中和平淡，入清则已不
见此类闲情逸致的诗文。

吴伟业（1609—1671），在《意
难忘·山家》词中写道："把瘿
尊茗碗，高话桑麻。穿池还种柳，
汲水自浇瓜。霜后橘，雨前茶。
这风味清佳。喜去年，山田大熟，
烂漫生涯。"

杜浚（1611—1687）在《茶丘铭》
中自称："吾之于茶也，性命之交也，
性也有命，命也有性也。天有寒暑，
地有险易，世有常变，遇有顺逆，
流坎之不齐，饥饱之不等，吾好茶
不改其度，清泉活水相依不舍，计
客中一切炎费，茶居其半，有绝粮
无绝茶也。"

范承谟（1624—1676），字
觐公，号螺山，辽宁省辽阳市人。
他是第一批降清的汉族文人范文
程之子。顺治间进士，先为内阁
侍读，后为浙江巡抚、福建总督。
耿精忠在福建反叛后，范承谟被

abruptly in this world, Coiled-Dragon-and-Phoenix Pie tea symbolizes wealth and rank." He also wrote "Inscription for Xu-Yuantan-preserved Poetry Volume on Tea Activity": "Mountains and streams conceal their figures, what I can only see is that the light wind takes flowers accompanied by tea smoke off the branch." Both poems expressed tea man's state of neutralization and insipidness while focus their attention on tea. There was no such poem or prose containing leisurely and carefree mood in the Qing Dynasty.

Wu Weiye (1609-1671) wrote,"Raise wooden cup and bowl to drink tea while talking happily about my plantation including willows crossing the pool, melon watered by pool water, after-frost tangerine and before-Grain-Rain tea both tasting better, good harvest of last year and this open-and-natural style life" in his poem "In-mountain Home".

Du Jun (1611-1687) said in his article "Tea Mound" that "The relation between me and tea is a fatal one, I belong to tea and tea belongs to my life. Season changes, road varies, thing differs, good luck goes with bad one, ups accompany downs, starvation comes after satisfaction, but my love for tea will remain as it was like running water from spring, tea fee amounts to half of my living expense, I can live some days without food but can not live a single day without tea."

Fan Chengmo (1624-1676), whose style name was Jingong and assumed name Luoshan, was born in Liaoyang City of Liaoning Province. He was the son of Han-nationality scholar Fan Wencheng who was one of those first yielded allegiance to the Qing Dynasty. He had been selected as one of the successful candidates in the highest imperial examination in the Shunzhi Reign, then document translator, provincial governor of Zhejiang, and governor general of Fujian Province. Fan Chengmo was captured after the Geng

抓，后被迫自杀。这是一位曾深受康熙帝器重，并在当时的官场有美名，在当时的浙江民众中有好口碑的官员，同时也是一位文人，著有《抚浙督闽稿》《吾庐存稿》《百苦吟》《画壁遗稿》等书。

这样一位官员兼文人，同时也是一位爱茶人，他所写的茶诗格调轻快、境界清新。从其所写的一些茶诗来看，作者常在轻快、清新脱俗的茶境中舒放性灵。如作者所写《柳茶》："竹炉烟里想山家，蟹眼烹来谷雨芽；一自掉船滩溜外，焙成柳叶当新茶。垂垂绿树即为家，未到清明摘嫩芽；浪说卢仝七碗茶，武彝梦断雨前茶。千泉泖水足山家，柏子初然试紫芽；我愧卢仝耽此癖，难途柳叶作新茶。"在《煮茗》诗中作者写道："旗枪泡露拂烟斜，汲井频将活火加；遥望福堂心供养，赵州一盏绿云茶。"

毛奇龄（1623—1713）《试茶歌》中记述了他的试茶：

"我来试茶值社后，少妇入云绿洗手；山头烂石膏沫多，雪砾霜崖绝枯朽；青丝笼子剪香叶，

Jingzhong's rebellion, later was forced to commit suicide. Fan Chengmo, who had been highly esteemed by emperor Kangxi, shared a good reputation in the officialdom and among the public in Zhejiang Province at that time. He was a government official and a literati as well whose works are as follows: Manuscript on My *Working Experiences in Zhejiang and Fujian, Manuscript Kept in My Cottage, On Sufferings, Posthumous Manuscript of On-wall Writings*, etc.

Fan Chengmo was also a tea man who created many tea poems which were relaxed and fresh in style. We could feel from his tea poems that Fan Chengmo usually released his natural disposition and intelligence in a light and refined tea-drinking atmosphere as he wrote in one of his poems "Willow Leaf Tea": "Bamboo smoke rising from stove reminds me of my in-mountain home, boil before-Grain-Rain bud tea till having crab-eye water bubble; slip ship off the beach to pick some willow leaves and wish to bake them as fresh tea. The willow leaves I picked are before Pure Bright Festival, What Lu Tong dreams to drink is before-Grain-Rain tea from Wuyi Mountain. Try to make tea with purple bud of cedar seed and spring water, I feel shamed when mentioned Lu Tong's name because of taking baked willow leaf as fresh tea." Fan Chengmo wrote in another poem "Stewing Tea" that: "Dew-wetted sign flag swinging makes the smoke oblique, dip water from the well and frequently add fuel into the fire; bless my good fortune with hearted worship accompanied with one cup of green cloud tea."

Mao Qiling (1623-1713) described his experience in his poem "Experiment on Making Tea": "I am trying to make tea, young woman climbs high mountain getting hands totally green; broken cliff on top of the mountain is covered with gypsum foam, dying trees stand among white gravel cliff; black

箸裹焙成卷银鼹；不需木棹共桑
砧，何用铜匙并铁铗；相携且试
耶水滨，青黄黑白甘苦辛；粗柑
细药似难较，鸡苏狗棘非其伦；
须臾地炉活火起，沸向花阶石萝
里；半杓疑分乳窟泉，满船刚载
南泠水；倾来清莹作冰雪，扫却
黄瓷细沤沫；分明眼底见幽兰，
骤使心中断消渴……"

袁枚（1716—1797）不但喜欢
家乡的龙井茶，对常州阳羡、湖南
洞庭君山也有很高评价。在他亲身
游历了武夷山，对当地饮茶方式有
所了解后，才彻底改变了对味苦如
药的武夷茶的看法。他意识到，以
前尝过的武夷茶或真假难辨，或水
脉不合；还在于，一地之茶，有一
地之独特烹饮妙法。

难能可贵的是，在茶人埋
头于形而下的品鉴时，袁枚提出
品茗的"味外味"。把品茗上
升为一种艺术；这与古代"字
外功""功夫在诗外"有相似之
处，提倡文化修养和品德培育。
这在整个清代并不多见。

乾隆皇帝作为一代君王，对茶
叶种植面积扩大，增加新的茶种和
加工方式，茶具创作方面，起到无

cage contains fragrant tea leaf, bake it wrapped by bamboo leaf until it turns into silver white; wooden tea-making utensils such as spoon and pliers instead of metal pieces should be used, tea leaves having different colors and various natures mix together with water, it seems difficult to tell the difference among them excluding Borneol Mint and Rhizoma Cibotii; fire from pit furnace boil the water in a short while, boiling water drops go even as far as Uanea longissima on the stone stairs; half scoop of water seems to be that from grotto spring actually Nanling water from Mount Jinshan; the clear and glistening Nanling water looks like icy snow when pouring into yellow porcelain tea ware sweeping away fine bubbles on the surface of tea ware; as if orchid appears in my eyes driving away the feeling of thirst instantly..."

Yuan Mei (1716-1797) liked not only Longjing tea from his hometown but also estimated highly Yangxian tea from Changzhou and Junshan tea from Dongting in Hunan. Yuan Mei's view towards medicine-like-bitter Wuyi tea was completely changed by his personal travel to Mount Wuyi after he saw in person the local way of drinking tea. He realized that Wuyi tea he once tasted might not be the genuine one or not be cooked in the unique local way.

What deserves praise was that Yuan Mei put forward his view "subtext of tea tasting" when people indulged in only tea-related activity. he believed that tea tasting was a kind of art which must contain cultural attainment and moral cultivation like what ancient Chinese sayings expressed "Extra self-cultivation beyond calligraphy" and "Effort beyond poems". This was rare throughout the Qing Dynasty.

Emperor Qianlong played an irreplaceable role in expanding plantation area of tea trees, cultivating new tea types, creating new processing methods and tea ware

可替代的作用；在引导社会品茗并使茶成为国饮方面功德无量。但我们必须清醒地看到，虽然他的茶诗数量颇丰，但多沦于清赏自娱，艺术性和思想性不高。这不仅与生活的锦衣玉食有关，与他们君臣歌舞升平沉溺奢华，回避深层社会矛盾也不无关系。

就茶道文化的高度而言，我们看到清代只注重瀹茶、鉴水、鉴茶的技巧而忽视对"和""雅"为核心理念的茶道文化的追求。其《观采茶作歌》有"雨前价贵雨后贱，民艰触目陈鸣镳""敝衣粝食曾不敷，龙团凤饼真无味"等关心民瘼疾苦的文字，却匆匆转移到前朝皇帝的龙团凤饼上去，其实是五十步笑百步。

与前朝历代官府一样，"催贡文移下官府，哪管山寒芽未吐，焙成粒粒比莲心，谁知侬比莲心苦"（陈章《采茶歌》）。其实即使在清代，"一瓯水白茶如雪，足抵人间七品家"（袁枚《湖上杂诗》之十八），上供炒青茶芽同样给茶农造成巨大负担。

就茶道精神而言，袁枚的"味外味"是有清一代的高标。金圣

production, also made a boundless contribution in guiding social drinking habit and turning tea into national drink. But we need to be clearly aware that although Emperor Qianlong composed a large quantity of poems, most of them aimed at appreciation and self−entertainment instead of artistic and ideological presentation. It was not only associated with his extravagant life, his own addiction to luxurious life with singing, dancing, etc., but also his unconscious attempt to avoid facing deep social contradictions.

It was obvious that tea men of the Qing Dynasty paid attention only to brewing tea, identifying water and tasting tea while ignored the pursuit of harmony and elegance which were the core concept of tea culture. Sentences in emperor Qianlong's poem "Watching Tea−leaves Picking" showed his concern about common people's hardships "The price of before−Grain−Rain tea is higher than after−Grain−Rain tea, I run into people's difficult life"; and "Ragged clothes and coarse food are even not sufficient for their daily life, 'Coiled−Dragon−and−Phoenix Pie' tea is really tasteless."Then, his concern was replaced by the comparison he made among imperial 'Coiled−Dragon−and−Phoenix Pie' teas of different dynasties. Royal court tea tribute was a huge burden for tea farmers of the Qing Dynasty like all previous dynasties just as Chen Zhang wrote in his poem "Tea−leaves Picking": "When the document on urging local officials to collect royal court tea tribute arrives, tea farmers are forced to pick tea bud in cold mountain wind, then to bake tea bud which finally turn to be as small as lotus nut, who knows that tea farmers' heart is as bitter as lotus nut." Also as Yuan Mei wrote in the eighteenth poem of *Miscellaneous Poems on the Lake*: "The cost of one cup of snow−white tea made of cloudy water equals to expense of a seventh rank family." Serving roasted green tea sprouts as

叹、钱谦益、曹雪芹、钱大昕、
龚自珍整个清代其他文化人，在
茶道思想茶道精神的认识和普及
上，都没有达到袁枚的高度。但
是清代的茶画却并不逊色于前代，
这也使我们感到一种精神慰藉，
因为画中的茶道精神是客观而直
接地展现在我们面前，为认识中
国茶道打开一道美丽的视窗。

tribute brought a great burden to the tea farmers.

As for the spirit of tea ceremony, Yuan Mei's view "Subtext of tea tasting" was a high standard for tea culture in the Qing Dynasty. The understanding on the spirit of tea ceremony of other intellectuals of the Qing Dynasty such as Jin Shengtan, Qian Qianyi, Cao Xueqin, Qian Daxin, and Gong Zizhen had not reached the height of that of Yuan Mei. However tea-related painting of the Qing Dynasty was not inferior to its kind of the former generations making us feel a bit of spiritual comfort and at the same time opening a beautiful window in front of us to know more about Chinese tea culture objectively and directly.

【注释】

[1]《满族研究》2013年第3期，第65页。

[2]《满族研究》2013年第3期，第66页。

[3]《农业考古》2014年第2期，第149页。

[4]《农业考古》2014年第2期，第152页。

[5]胡长春:《陆廷灿〈续茶经〉述论》，《农业考古》2006年第2期，第262页。

[6]爱新觉罗·弘历:《观采茶作歌》，选自刘枫《历代茶诗选注》，中央文献出版社，2009年，第193页。

[7][8]陈章：《采茶歌》

【Notes】

[1] *Studies on Manchu Nationality*, No.3, 2013, p.65.

[2] *Studies on Manchu Nationality*, No.3, 2013, p.66.

[3] *Agricultural Archaeology*, No.2, 2014, p.149.

[4] *Agricultural Archaeology*, No.2, 2014, p.152.

[5] Hu Changchun, "Discussion on *Supplement of Tea Classic* by Lu Tingcan", *Agricultural Archaeology*, No.2, 2006, p.262.

[6] Emperor Qianlong, "Watching Tea Picking and Singing", in Liu Feng, *Notes on Selected Poems about Tea in Past Dynasties*, Beijing: Central Literature Publishing House, 2009, p.193.

[7]Chen Zhang, "Song of Picking Tea".

[8]Chen Zhang, "Song of Picking Tea".

图 275

清·萧云从（1596—1673）·《关山行旅图》·重庆三峡博物馆藏品

原画高 26.6 厘米，宽 489.5 厘米。

崇山峻岭之中，浓荫蔽日之下，飞流涌雪之旁，有一处亭阁，有一个茶室，有一个事茶男子，是何方神圣至此，静静地叙话品茗？

图 276

清·萧云从·《纳凉竹下》·北京故宫博物院藏品

原画高 22.8 厘米，宽 15 厘米。

题款："纳凉竹下，仿范中立意也。"古人也从经验中悟得：饮用绿茶可以防暑降温。

图 277

清·萧云从·《石磴摊书图》·北京荣宝斋藏品

　　1669 年绘，原画高 132 厘米，宽 66 厘米。

　　这里读书、论诗、品茗的幽境引人向往。

图 278

清·顾见龙（1606—1687）·《秋闲论古图》

原画高 140 厘米，宽 68.5 厘米。

图中只有烹茶涤器的两个书童外，其他人都给人一种收缩畏寒的感觉。虽然秋色浓艳，似乎寒

意已来。三位文士面对桌上的古画已经到了忘我境地。

图 279

清·严湛·《树下对饮图》·英国伦敦莫斯公司藏品

此画 1663 年画。

严湛，生卒年不详，为浙江绍兴人，师承陈洪绶，画作与其师风格相近。端
杯对茶，倾注了恭敬与珍惜之情。此长安当是我们现在的西安。

图 280

清·戴本孝（1621—1691）·《平台幽兴》册页·北京故宫博物院藏品

1690 年制，高 21.8 厘米，宽 16.4 厘米。

画中碾槽、风炉和茶壶、茶杯，表明在林泉高致、平湖潋滟的环境中，茶带给人

精神愉悦与幽静。

图 281

清·梅清（1623—1697）·《瞿硎石室图》·北京故宫博物院藏品

画高 166.5 厘米，宽 63 厘米。

茶禅一味，还是物我两忘。

图 282

清·周洽（1625—1700）·《竹溪春昼图》·西安美术学院藏品

画高 208.5 厘米，宽 1118 厘米。

周洽，上海人。

文人在山间亭子竹林相聚，有僮仆带来茶器，连烹茶之水也是专门预备的。

图283

清·吕焕成（1630—1705）·《蕉阴品茶图》·西安美术学院藏品

吕焕成，浙江余姚人。清初"吴门画派"代表之一。《蕉阴品茶图》高427厘米，宽92厘米。就茶道文化研究而言，这是一幅难得一见的真正意义上的品茶图。在蕉树林下，主人一人坐于石桌之旁的披着红色坐垫的石头之上；右手用拇指、食指、中指端持白瓷杯，无名指挨着杯壁，小指如女士莲花指翘起，左手半握拳，食指弯曲前指，正在向奉茶的弟子讲评前三种茶的优劣得失；奉茶弟子正在与主人搭话，小心谨慎地看着主人的脸色，手捧紫砂壶，上身前倾，抬头看着主人。主人或师傅，在对茶叶本身滋味品鉴的同时，自然会说到侯汤泡茶时手法上的得失，因而负责泡茶的弟子紧张而又忐忑不安的神情十分明显。半跪半坐于地面上的童子自然听到师傅的评点。茶汤是否成功十之八在水，而煮水到什么样的程度，是他的责任。师傅在言谈间自然涉及他的责任。他认认真真地用蒲扇扇火，聚精会神地注意炉上壶水的水温变化。四种茶壶，两边包括弟子手中的是两个紫砂壶，色泽一浅一深，一个白瓷执壶，一个褐色壶。从形制看，可能是锡壶。喝茶的杯子两套各两个。从蕉叶和地上小花烂漫的情形以及师徒三人的衣着看，时序大约在清明之后，蕉树尚未开花。炉子、奉茶弟子手中的壶与师傅手中的杯，恰好在一条线上，这也暗合了三者之间的内在逻辑关系。一人专门负责烧火，一人负责冲泡，师傅专门品鉴。

品茶有几种方式，或者同一种茶叶用不同的壶，用同火候的水来冲瀹；或同一种茶，因壶不同，因而用不同火候的水来泡。也可以用同一种茶在同壶中就长短不同的发泡的时间进行品评。

在画中，我们看到了现代人称为茶海的古匜形茶器。该画表明，明末清初，中国茶人对茶器的追求、对茶品品质的品鉴，十分讲究。

图 284

清·王翚（1632—1717）·《一梧轩图轴》·台北故宫博物院藏品

本幅纸本，浅设色画。高104.1厘米，宽54.2厘米。

梧桐高耸，草轩简约而敞阔，主人放下茶杯后即拨琴发声，仙鹤即将鼓翅起舞，随音律而动，侍者站在侧后。湖石环列，芦苇竹子相间。其场景，将琴、茶与周围环境很好地融合起来。题记表明此作是作者42岁（1673）时的作品。他由吴中游淮扬，与好友查士标聚首累月，查有诗赋相赠。

图 285

清·王翚·《临许道宁山水轴》

本幅绢本，设色画。高 109.9 厘米，宽 57.4 厘米。

康熙戊午（1678）夏，王翚过金陵，访笪重光，见到许道宁画作，遂有此仿作。画中山秀，云浓，高泉下泻，湖平，孤船轻渡，楼阁半隐于树木之中，三两好友，云中漫步。舍中有茶否，品茶论诗，何等高雅。不见茶，不见器，但这是理想的茶道境界。（本图摘自台北《故宫博物院画系》）

图 286

清·王翚·《石泉试茗轴》·台北故宫博物院藏品

　　所谓试茗，也是在鉴水，画面三组人物：下面一人独坐亭子中，有茶壶在旁，

中间一人正策杖走过石桥，上面一人静坐草房之中看僮仆扇火煮茶。

图 287

清·王翚·《刘松年竹园逢僧图》·台北故宫博物院藏品

　　原画高 28.3 厘米，宽 44.5 厘米。

　　刘松年乃前朝赵宋时画家，对茶事极为娴熟。以其掌故创作，其实已经预设了画作的情调：茶、禅在小桥流水之旁，茂林修竹之边，古松之下，是茶与禅共理的别样意境！

图 288

清·吴历（1632—?）·《云白山青图》·台北故宫博物院藏品

　　此画作于作者37岁（1668年）。作者题记："雨歇遥天海气腥，树连僧屋雁连汀。松风谡谡行人少，云白山青冷画屏。戊申九月六日余从毗陵归虞山，风雨寂寥中有啜茗焚香之乐。八日晚晴。喜而图此，吴历并题。"钤"吴历"朱文印。啜茗之后有佳作，此为证也。毗陵，今江苏常州市。

图 289

清·担当·《行旅图轴》（局部）·云南省博物馆藏品

原画高 108 厘米，宽 29 厘米。

行旅艰辛，茶炉茗碗不可缺少！

图 290

清·担当·《山水图》（局部）·北京故宫博物院藏品

　　原画高 25.2 厘米，宽 24.9 厘米。

　　高山流瀑，泉水清澈，用炉煮茶，一大快事。你们车马炮厮杀，
让我再喝一杯茶。

图291

清·高简（1634—1797）·《静坐啜茶图》·天津博物馆藏品

  画高74.2厘米，宽46.5厘米。

  清明节最先是二十四节气之一，唐中期演化为祭祀节日，晚唐宫廷有了清明茶宴。明前茶一直是国人热捧的茶中珍品。作者清明日题画也许出于偶然，但清明节追茶一直是民族风尚。这是一个极富特色的私人茶室，当作都篮的是一个宽敞的两层柜台，放于茶室外屋檐下。

图 292

清·石涛(1642—1707)·《看山图》·上海博物馆藏品

　　画高 129.6 厘米，宽 54.3 厘米。

　　作者题记表明，其画作当事人未必重视，但索要收藏者却都是真心实意爱他画的人。欣赏揣摩画作意趣时，焚香沐手，以茶清心，十分虔诚！

石涛·《看山图》（局部）

图 293

清·石涛·《狂壑清岚图》·南京博物院藏品

　　原画高 1649 厘米，宽 559 厘米。

　　瀑布飞流，岚风狂作，与屋下品茗叙话的从容宁静形成鲜明对照。因为茶助而文思清而心绪宁。

图 294

清·杨晋（1644—1728）·《豪家佚乐图》（局部）

　　绢本设色，高 562 厘米，宽 1274 厘米。

　　杨晋，字子和，一字子鹤，号西亭，自号谷林樵客、鹤道人，又署野鹤，江苏常熟人。山水为王翚入室弟子。《豪家佚乐图》实际上给我们展现了一幅南方豪门大族妇女的品茶生活。画幅左端一妇女挥扇烧水，一个小孩子挥扇追蝶，表示时间在孟春。整幅画的主体是着浅色裙襦的两位夫人品茶叙说。

杨晋·《豪家佚乐图》（局部）

　　画幅右端另一个侍女捧着紫砂壶、白瓷杯在急急地行走。
这是夏日生活的一个场景。

图 295

清·黄应谌·《陋室铭图》·台北故宫博物院藏品

　　黄应谌，生卒年不详，原画高 243.3 厘米，宽 158 厘米。

　　琴童抱琴肃立，书童奉函恭候，一个茶童端水，一个茶
童生火。庭内有香炉、珊瑚树、梅花盆景。室内一个展开诗文，
急诵缓吟，其他三人躬身谛听。

　　茶坊紧邻会客厅，有专人泡茶。

图296

清·吴宏（生卒年不详）·《柘
溪草堂图轴》·南京博物院藏品

画高1608厘米，宽798厘米。

作于康熙十一年（1672）。自
题"柘溪草堂图，壬子秋九月，拟
李咸熙笔意"。落款"金溪吴宏"，
钤"吴宏私印""远度"白朱文二方印。
诗堂及裱边有查士标、梁清标题诗。
朱彝尊、程远等七人题词。画中人
物在房舍内品茗论道。

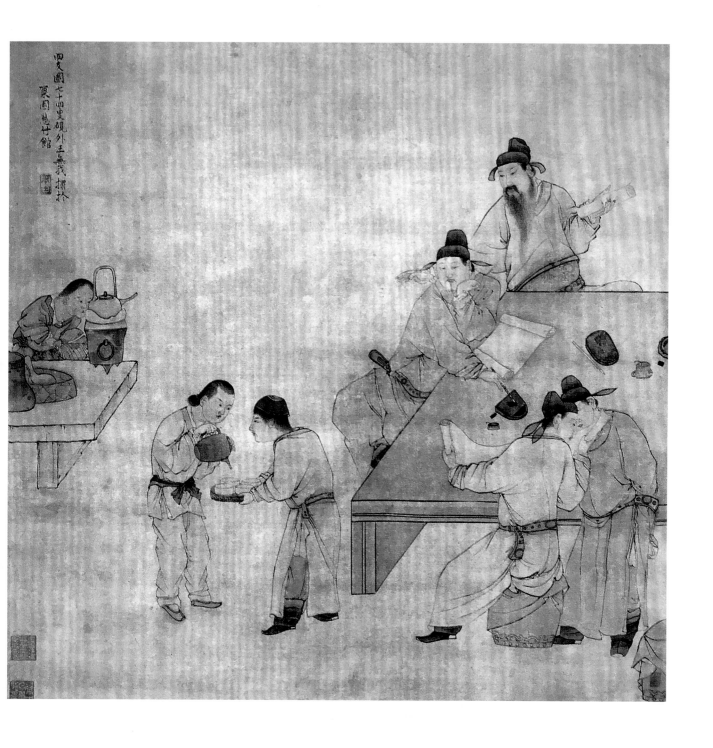

图 297

清·王树榖（1649—?）·《四友图轴》·北京故宫博物院藏品

原画高 66.3 厘米，宽 66.6 厘米。

可以看到，文人诗书绘画不离茶。三个书童烧水、冲泡、奉送，各司茶事，忙而有序。而文人们似乎脑子有点僵，等着一杯茶来浇开诗思。

図中の題記：

文章四友圖
古人文字偏珍惜
我亦如今惜墨金
慚愧老年誰為國
徽為君忘日不
塗鴉
康熙六十年秋七月
王慈竹山於桃軒題
於右松堂

图 298

清·王树榖·《文章四友图》·西泠印社藏品

1721 年绘，原画高185.5 厘米，宽 100 厘米。

图中五个书童，有三个与茶有关，一个提篮子；最前面一人抱一个囊团：有可能是名香及香炉，可能是茶笼子，如赵孟頫《写经换茶图》中的茶笼子，也可能是用布包裹的文章书稿。题记末句有"为君无日不涂鸦"，但桌面上未见书函，趴案者手中也只有薄薄几页，谈不上"无日不涂鸦"。

图 299

清·王云（1652—1735）·《山水图之七》·天津博物馆藏品

原画高 33.1 厘米，宽 49.3 厘米。

图中似为一对老年夫妇黑白对弈，书童在另一个山洞里扇火煮茶。他们头顶雨雾缭绕，山石苍茫，脚边溪流淙淙。这是恬淡宁静、自由自在的生活。茶，增益了清寂，也映衬了这种闲舒。

王云·《山水图之七》（局部）

图 300

清·萧晨（1656—?）·《课茶图》·北京故宫博物院藏品

　　画高 19 厘米，宽 24 厘米。

　　"课茶声细炉中雨"，以雨声来比喻煮水时微微作响，犹如潇潇雨下。课就是
一种有技艺的、有专业修养的烹茶过程，绝非随意而为!

图 301

清·袁江（1662—1735）·《米家书画船》·北京故宫博物院藏品

　　原绢画高 38.6 厘米，宽 32 厘米。

　　图名《米家书画船》，典出宋黄庭坚《戏赠米元章》诗之一："沧江尽夜虹贯月，定是米家书画舫。"
米家船后来成为"米家书画"代名词。任渊注："崇宁间（1102—1106），元章为江淮发运，揭牌于行舸之上
曰'米家书画舫'。"金代元好问《钱过庭烟溪独钓图》诗之二："小景风流二百年，典刑来自米家船。"
明王时敏《题自画关使君袁环中》："割取一峰深秀色，可堪移入米家船。"可见米家画的生动景象。这是
一幅诗意画。米芾为徽宗朝受重用的书法绘画大家与苏轼黄庭坚蔡襄（一曰京）并称书坛"苏黄米蔡"。我
们也由此可知，米芾喜在船中创作。他作画时，船夫躲在船头钓鱼，侍者在船尾生火煮茶，三人各行其是，
互不干扰，真是人间天堂。米芾《苕溪诗帖》："楼阁明丹垩，杉松振老髯，僧迎方拥帚，茶细旋探檐。"

图 302

清 · 朱珏 · 《德星聚图》· 北京故宫博物院藏品

原画高112.2厘米,宽50.5厘米。

德星,指贤士。德星聚,意为两位(以上)贤人相聚。一位贤者率领众子弟来登门拜访另一位贤者。客人坐于三足墩凳上,众弟子簇拥站立于后;主人坐于床榻之上。两位贤者啜茶交谈。主人书童正在摇扇煮水,为其他来访者备茶。贤者近旁各有一位活波可爱的小孩依偎在身边。两位小孩既拘谨又兴奋的情态极为明显。主人后面两位年纪稍大者恭立其后,似在作揖行礼。另有两人在主人旁边的桌案上摆置饭碗。整个画面人物较多但却极为协调、宁静、和雅。

一勺金碧霄活火真嫩
乳中肴萬山精白石泉源
養　　坚天池上昴寺僧試茶

图 303

清 · 高凤翰 · 《天池试茶图》· 沈阳故宫博物院藏品

　　作者 1733 年绘，高 28 厘米，宽 34 厘米。

　　天池，指山上面的小潭、小池。从作者落款（法阜）可知，是与僧人（近旁寺院）进行瀹茶品泉。不排除在东北长白山天池品茗的可能性。

图 304

清·丁观鹏（? ——1771 或 1736——1795）·《太平春市图》·台北故宫博物院藏品

作者 1742 年画，原画高 30.5 厘米，宽 235.5 厘米。

春天集市上有售卖时果、茶水者，也有自带瓜形紫砂壶品饮闲聊者，有敲锣唱戏者。

图 305

清·丁观鹏·《乞巧图》（局部）·上海博物馆藏品

原画高 28.7 厘米，宽 38.45 厘米。

乞巧，是七夕节最普遍的习俗，就是少女们在七月初七的夜晚进行的各种乞巧活动。东晋葛洪的《西京杂记》有"汉彩女常以七月七日穿七孔针于开襟楼，人俱习之"的记载。据《开元天宝遗事》载，唐玄宗与妃子每逢七夕在清宫夜宴，宫女们各自乞巧，这一习俗在中国民间也经久不衰，代代延续。

穿针引线、绣花织锦是古代女子智慧的集中表现。乞巧，从字看就是在七月七日七夕日乞求织女赐予智慧，后来又引申为乞求理想的婚配、幸福的生活。凡女子所能展示的才艺都会搬到七夕来比试。图中所绘是女子们各自携带茶壶（茶叶）进行斗茶比赛。这幅图是女性进行茶艺比赛的最早的艺术表现。

图 306

清·丁观鹏·《仿仇英西园雅集图》

　　图的左上角表现的是备茶情景。将十六人分为四、九、三人三组外，又有备茶的侍者二人和奉茶的
仕女二人。

呼童收曉露煮
茗付都籃英嫩
春花綴珠螢夜
氣含銅仙高漫
挹荷蓋味宜探
活火調蒲簽清
聲溢鼎甘

图 307

清·黄钺徵·《黄钺徵符康阜册·晓露烹茶图》·台北故宫博物院藏品

　　黄钺徵，生卒年不详。图中题记云："呼童收晓露，煮茗付都籃。英嫩春花缀，珠萤夜气含。铜仙高漫挹，荷盖味宜探。活火调蒲簽，清声益鼎甘。"钤朱文印"漱润"，题名"晓露烹茶"。

　　诗中"铜仙"典出汉武帝。其为求长生不老，下令在长安建章宫内建造神明台，上面再铸造铜仙人双手捧铜盘，以此来求得仙露。唐经学家颜师古注《汉书·郊祀志》："《三辅故事》云：建章宫承露盘高二十丈，大七围，以铜为之，上有仙人掌承露，和玉屑饮之。盖张衡西京赋所云'立修茎之仙掌，承云表之清露，屑琼蕊以朝餐，必性命之可度'也。"这里所收之露不用铜人巨掌，而是荷叶。瀹茶以水质为主要因素，以荷露烹茶可谓别出心裁。

图 308

清·陈枚（？—1864）·《人物册》·北京故宫博物院藏品

　　陈枚，上海人，所绘《月曼清游图》之《秋桐聚戏图》，描述贵族妇女在秋桐下聊天与赏茶的场面。

我们可以看到白色茶盏中的汤色略呈黄色——古人所谓"缃色"。

图 309

清·焦秉贞·《历朝贤后故事图》之《孝事周姜》·北京故宫博物院藏品

焦秉贞，生卒年不详，康熙朝宫廷画家。虽然两周时期还没有华丽的锦绣细瓷，还没有带托茶盏，但是奉汤馔孝敬长辈的观念和习俗自古就有。至少在画家所在的清朝，中国孝事先祖长辈的传统依然占据人文伦理的主导地位。而茶在人们的日常生活和礼仪制度中具有极为崇高的地位，特别在与妇女有关或妇女为主体的活动中，更是如此。

图 310

清·吕时敏·《茗情琴意图》（局部）·青岛博物馆藏品

《茗情琴意图》图题记云："康熙丁丑茗川吕子时敏为余写此小照。距今八年矣。于时拂髯青丝萦腰茜绶，书成盐铁，
最嗤桑孔之谋。节驻钱塘雅慕白苏之绩。当公余而寂处，每兴发以幽寻。孙子荆癖染青霞；仲长统情耽白水。桐岗竹屿，
是处行窝（卧），鹤子凫翁，多称密契，爱待玉琴于石畔，且呼香碗于花间床余，壮武以编牙签，略识砌长，康成之草

堪娱，拈秃笔以长谣，一任批风抹月，策孤节而远眺，何妨涉礀穿云，追溯华年，倍多乐事，攸过无闻之岁，重寻有美之堂，吏牍台符漫堆，琴几尘容，俗状难对，青山顾宦，况其如冰愧！君恩之似海，空餐可鄙，但期无添职司，胜践犹存，未忍递忘林壑。

　　览是图也增吾感焉　乙酉春日　北轩主人　自题

　　钤"郎廷极印"，朱文，"紫衡"朱文。

　　画末端系吕时敏题款，系绘画作者。

　　画末端作者题款：

　　"北轩先生郎公台阁，人宗当代。仰之如景星庆云。顾性爱闲旷，时蒞两浙。公务之暇辄觞咏于六桥三竺间。余亲炙其逸性高致，而作是图。窃以　公将来位秩益崇，经纶猷业，清暇常稀，恐未能从事登涉也。姑留此为异日平泉绿野之式。茗水吕学拜识。"

　　钤朱文"吕""学"二印

　　丁丑年当为1697年。

　　画中人物郎廷极自题款（相当于序文）之后第一组画是两个书童行走在小桥流水前的竹林、土丘上，向画面左方行走。小花灿灿，小河奔流。幼小一个背竹笔帘卷成五轴书画包囊，个子大一点的抱琴囊。

　　第二组画是整画核心。郎廷极（1663年—1715）衣青衫，赤脚，席地而坐于河岸石上，这块石头延伸至水面之上，下空，半开的蓝皮书卷放在右手跟前。一长一少的两个侍者正在整理石床：长者正在用布巾擦拭。年少者端盖碗白瓷杯。石床上有风炉、汝窑水钵、颜色一深一浅的紫砂壶、用纱网罩盖着的食物水果、寿桃（馒头）、白釉茶托子（与少年手上的配套）。石床安置于松下大石之旁。三棵松树苍劲古雅。主人看着潺潺流水，悠然自得。画中的郎廷极是35岁的相貌。方脸庞，白皮肤，留胡须。

　　第三组画是一光头少年右手提一梯形紫砂水壶，左手捧端圆茶盘，茶盘里一小紫砂壶、一龙泉窑茶盏。边走边看小径边的紫色野花。他朝画面右方行走。三只鸟儿在水中的石头上鸣叫，一只野鸭在水中游走。

　　第四组画是一个书童在邻水的亭轩里整理书房，青桐钫里插着两朵大叶白色花束，绿叶生机勃勃。书童正在给香炉里添加香末。连接听轩与石山的拱形桥下的水流相当湍急。

　　无疑，前三幅是相互联系的整体。第四幅也必然是组画的一部分：品茗抚琴后，还要回到亭轩书房，进行书画创作。

　　《郎氏宗谱》载，自明末至清光绪年间始祖郎玉以下十代世系。郎廷极，清代隶汉军镶黄旗，奉天广宁（今辽宁北镇）人。湖南布政使、山东巡抚郎永清子。康熙间以门荫授江宁府同知，迁云南顺宁知府，累擢江西巡抚，督造官窑瓷器，世称郎窑，官终漕运总督。

　　在整幅画中，我们没有看到在历史上极负盛名的郎式釉里红瓷器。也许督造瓷器在后。

图311

清·髡残（1612—1692）·《山水图》

髡残俗姓刘，出家为僧，名髡残，字介丘，号石奚，白秃，石道人，残道者，湖广武陵（今湖南省常德）人。喜画山水，师法王蒙。其作品苍浑茂密，意境幽深。将朝代更迭、个人经历，挥洒于笔墨之中。本图在雄浑苍茫色彩斑斓中，又给人另一个境界：万丈山水之外，是平和清淡的品茗谈玄。他与石涛合称"二石"，又与朱耷、弘仁、石涛合称"清初四画僧"。

图 312

清 · 郎世宁（1688—1766）·《弘历观画图像》· 北京故宫博物院藏品

纸本设色，高 1364 厘米，宽 62 厘米。

此图绘乾隆皇帝弘历坐在浓荫下凝神聚气观看丁云鹏所画的《洗象图》。画中人物衣纹用笔也似丁氏之《洗象图》一样用流水描，画里画外，达到统一一致。人物脸部用西洋的明暗凹凸法来表现，很有立体感。

我们选此图作为茶道图，是因其桌案上有红蓝两种紫砂杯，有笔架花瓶，实际是品茗观画，是古代常见的茶事活动。

图 313

清·郎世宁·《博古架前》·民间收藏

　　瓷版画

图 314

清·郎世宁·《雍亲王行乐·临轩品茗图》·民间收藏

　　瓷版画

图 315

清·郎世宁·《仕女奉茶图》·民间收藏

　　瓷版画

图 316

清·郎世宁·《十八学士图》之一·民间收藏

　　瓷版画

图 317

清·郎世宁·《十八学士图》之二·民间藏品

瓷版画

图 318

清·郎世宁·《衣裘照镜》·民间藏品

瓷版画

图 319

清·郎世宁·《雍亲王十二美人图》·民间藏品

瓷版画

图 320

清 · 王大凡 ·《桃李夜宴图》· 民间藏品

瓷版画

图 321

清·姚文瀚（18 世纪）·《四序图》之《游湖》·北京故宫博物院藏品

　　整图为绢本设色，高 31.5 厘米，宽 318 厘米。

　　描绘宫廷仕女春游、纳凉、游湖、赏雪四季不同之悠闲生活，构思巧妙，意境幽曲，刻画细腻，形象生动，笔墨工整，设色典雅。雅园之内，高阁之上，有人奉茶，定为皇室贵族。此作人物虽小，然曲尽姿态，形神兼备，款署"臣姚文瀚恭画"。本图为其中《游湖》，图中以泛舟采莲为主，年长的夫人在轩中观看为辅，仕女为她们奉茶。

图 322

清·陈字（1634—？）·《青山红树图轴》·北京故宫博物院藏品

泛舟于山水之间，携茶铛以品水，为文人雅好。右上朱文篆体方印"江左"题记："辛未仲夏之望，为子端表兄鉴定　枫轳弟宇（宁）小莲"。白文方印"陈字""小莲"。

图 323

清·冷枚（约 1669—1742）·《刻丝夜宴桃李园》·辽宁博物馆藏品

画高 136 厘米，宽 71 厘米。

原为廷宫制造，八国联军 1860 年入侵北京，流出宫外，为王伯珩所收，后为朱启钤所得《存素堂丝绣录》有记载。画中院落依山傍水而建，桃红李白，生机盎然，张灯结彩，一派祥和，茶酒间进，客主兴致高昂。白瓷茶杯、紫砂执壶与束口大腹酒缸同时置于李子树下的大方石桌之上，层层叠叠的食盒紧靠石桌。伸进水中的石头上有汤羹鼎、勺以及香炉、砚台、书画囊筒等，还有打开待写诗作画的纸张。原画为冷枚所作。冷枚，字吉臣，号金门画史，康熙后期为宫廷画师。

图 324

清·冷枚·《月下听泉烹茗图》·北京故宫博物院藏品

原画高 28.4 厘米，宽 30.9 厘米。

细流高下，淙淙有声；月映泉水，随波不定；与僧对谈，禅机妙应；生火吹鼎，

茶香袅袅！

图 325

清·冷枚·《邀月图》·西安美术学院藏品

　　本画 1739 年作。主人在秋日之夜携二书童，邀月品茗。月夜朦胧，桂树花蕾粲然，高台之上，三人情态各异。主人沉稳内敛而又舒缓自然，抬头望月，作无限遐想。奉茶童子双手端盘，内有白色瓷茶杯和茶点，恭敬谨慎。另一童子则袒胸，仰首牛饮先生执壶里的茶水。题图和落款为"邀月图　乙未仲夏写金门画史冷枚"钤"冷枚"白文方印和"吉臣"朱文方印。

图 326

清·冷枚·《赏秋图》·台
北故宫博物院藏品

　　原画高 119.8 厘米，宽
61.2 厘米。

图 327

清·冷枚·《十宫词图》

　　一番游船赏景，乱了乌发；将回轩亭，且将发花整理，清茶一杯，
滋润喉吻，见了君王，还将咏诗诵词。

图 328

清·黄慎(1687—1768)·《春夜宴桃李园图》·泰州市博物馆藏品

绢本设色。高121厘米，宽163厘米。

画中的人物衣纹作游丝描、铁线描，或连勾带染，挺劲放纵；以草书之笔入画，极具功力。作者虽师法陈老莲、上官周，然又自创新意。此图取材于李白的《春夜宴桃李园》诗意。人物情态各异，动静有别，生动传神。桌上有高提梁紫砂壶、鼎形香炉、茶杯、酒盅，一人左手端茶杯，若有所思。

图 329

清·蒋嵩·《渔舟读书
图轴》

蒋嵩，号三松，三松
居士，江苏江宁（今南京）
人，生卒年不详。读书湖
中船上，茶器相伴。

图330

清·华嵒（1682—1756）·《金
谷园图轴》·上海博物馆藏品

画高148.9厘米、宽94.1厘米。

华嵒，字德嵩，更字秋岳，
又空尘，号白沙道人、新罗山
人、东园生、布衣生、离垢居士
等，老年自喻"飘篷者"，福建
上杭人，后寓扬州。诗书画兼
具，时称"三绝"。

此图取材于西晋时曾任荆州
刺史的石崇在所营建的金谷园内，
坐听侍姬绿珠吹箫的故事。

图中三位书童，分别拿蒲扇、
端书画册函、捧端茶壶。两仕女在
绿珠之后，一人持团扇，一人端白
瓷碗，碗中有勺子。石崇（明代打
扮），坐于圆形地毯之上，前有香炉、
白瓷茶杯，左倾卧靠隐几。题款："金
谷园图　　壬子小春写于研香馆之
东窗　新罗山人　嵒"。钤白文印：
"东园"。壬子当为1732年，为
华嵒71岁时作品，是他的代表作
之一。

华喦·《金谷园图轴》（局部）·上海博物馆

图 331

清·华嵒·《金屋春深图》·广
东省博物馆藏品

原画高 119 厘米，宽 57 厘米。

"乙卯夏日写于帘屋，新罗山人
华嵒"时在 1735 年，画家 74 岁。

图 332

清·华嵒·《梅月琴茶图》·上海中国画院藏品

　　原画高 49 厘米，宽 30 厘米。

　　寒梅、孤月，一杯清茶，枕衾孤单，以琴解离愁。乡愁、相思时以茶来平复心绪，舒解块垒，是中国人赋予茶的又一内涵。

複閣逗春寒欲寢枕衾單奈此梅梢月發我清琴彈
新羅山人寫於吟月樓

图 333

清·华嵒·《宋儒诗意图》·苏州博物馆藏品

　　原画高 86.9 厘米，宽 117.2 厘米。

　　画中题记："秋风万爽天气凉，此日何日升高堂……儿前再拜谢阿娘，自古作善天降祥。"落款："甲辰元月写宋儒诗意　新罗山人 华嵒"。高堂在上，献食、奉茶。时在 1724 年，画家 63 岁。

图 334

清·倪骧、张宗苍（1686—1756）·《黄鼎像》（局部）·天津博物馆藏品

原画高 62.2 厘米，宽 101.6 厘米。

此图表明当时人们使用锡罐储存茶叶。

图 335

清·李鱓（1686—1762）·《梅花图》·扬州
博物馆藏品

画高 58.9 厘米，宽 23.6 厘米。

画中写诗，以煮茶为题："疏篱矮屋傍溪沙，
桥外梅开一树花，渗入雪中分不出，扫来煮就小
春茶。"梅花和雪煮茶，当有别样滋味与芬芳！

图 336

清·金农（1687—1763）·《绿窗贫女之图》·金万戈藏品

　　人有贫富，茶无贵贱。纳日月灵气，蕴山水精华，一缕清香留人间，是茶的精神。

图 337

清·金农·《玉川先生煎茶图》·北京故宫博物院藏品

1759 年绘，高 24.3 厘米，宽 31 厘米。

金农，字寿门、司农、吉金，号冬心，又号稽留山民、曲江外史、昔耶居士等。别号很多，有金牛、百二砚田富翁等。为"扬州八怪"之首。金农自述："家有田几棱，屋数区，在钱塘江上，中为书堂，面江背山，江之外又山无穷。"

金农天资聪颖，早年读书于学者何焯家，与"西泠八家"之一的丁敬比邻。入都应试未中，郁郁不得志，遂周游四方，走齐、鲁，赴燕、赵；西去秦、晋，又至吴、粤。年方五十，开始学画，晚寓扬州卖书画以自给。妻亡无子，四处漂泊。

本图名《玉川先生煎茶图》，以其颇具特色的爨体书法题款。茶道亚圣卢仝，一生孤苦，此画可视为一种怀念。画中玉川子自煎茶汤，细细观看老妪用罐汲水。给人以消散、平淡的气息。

图 338

清·喻冲·《春庭茶乐图》·南京艺术学院藏品

原画高 133 厘米，宽 75.4 厘米。

画中牡丹花开，古树生机焕发。春茶新到，官人、仕女喜气洋洋，侍女们匆匆忙忙。

图 339

清·李方膺（1695—1755）·《梅兰图》·浙江
博物馆藏品

　　高 127.2 厘米，宽 46.7 厘米。

　　此图虽是壶、杯、花为主的静物，但书法内容则
告诉我们：壶中放置砂石有利于净化存贮之水。

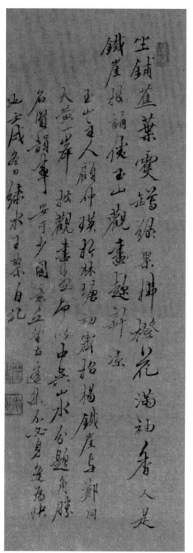

图340

清·王概·《玉山观画图轴》·北京故宫博物院藏品

王概（生卒年未详），康熙时代人，字东郭，又字安节，浙江秀水（今嘉兴）人。一直以来居住于江宁（今江苏南京），与兄王蓍、弟王臬是最早编订中国画技法画谱《芥子园画谱》（1679年套彩印刷）的画家。其大幅山水图轴闻名天下。这幅《玉山观画图》是其代表作之一。自题款为"壬戌冬日绣水王槩自记"。时间当在（1682年）康熙二十一年。王概与当时名流汤燕生、李渔、程邃、孔尚任、周亮工等交往，亦可作为考察其年龄生卒的参考。

画中主体画面是三位文人共同盯着画册，每人各有一白瓷茶杯。旁边一小孩，衬托出三人极为认真，不知道看什么宝贝，他也来凑热闹；后边的仆人趴在主人的椅背上，已经瞌睡，表现他们三人已经看了好长时间！另一桌上两人，旁边一白衣侍者右手端着几个空茶杯向天上张望。

图 341

清·王概·《秋山喜客图》·浙江省博物馆藏品

　　作者 1682 年作，画名为《秋山喜客图》可作二解。一是秋山美景苍茫，吸纳天下来客。"秋山喜客未草草，
客来亦谓此卷好！""山童收书为迎客，稚童扫叶供试茶。"客主二友对坐，一壶龙井，不管他秦汉魏晋。二是
客喜秋山。客居秋山，赏枫叶红了，看山云缭绕，听流水击石"飞泉推雪下青天"，如雷如天鼓。苍苍秋山助画

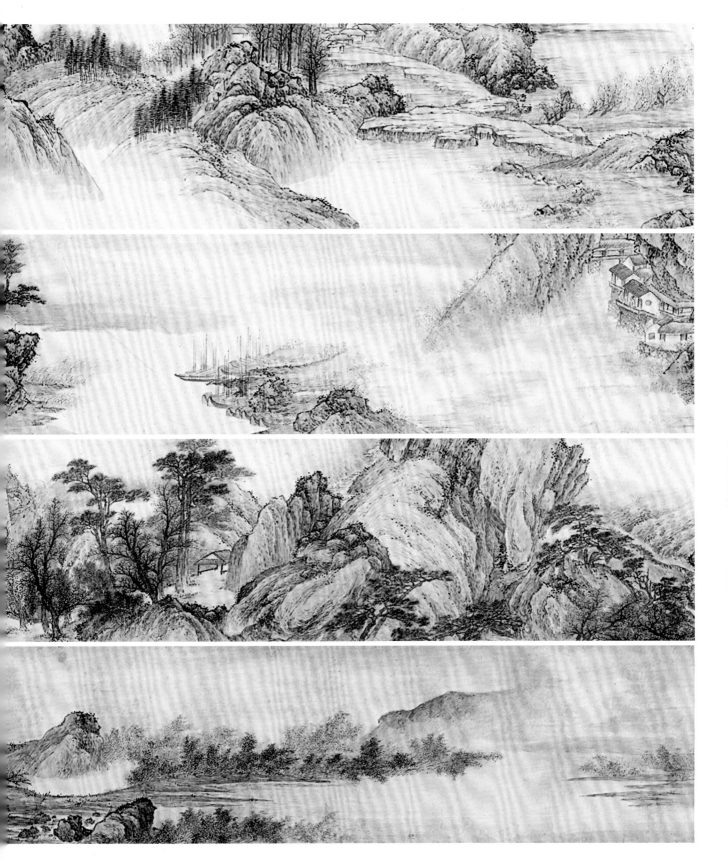

心："问余结撰为何人，我道与君交有神。平生爱画入骨髓，数遣书问何殷勤？于时龚开闭户写，名茶更寄庙后香，藏墨叠拣金壶液。"

对这幅气势磅礴的山水画，画家也是极为满意的："画成高卧霜正寒，十日五日且自看。仿佛开轩走麋鹿，沉溎入夜闻波澜。"而且骄傲地认为："莫谓此前吴徐笔墨不易得，但为身逢许苍雪。"而画为同年许苍雪之父所作，时在 1682 年十月，与前画《玉山观画图》创作于同一年的秋杪十月。

王概·《秋山喜客图》（局部）

图 342

清·陈枚（？—1684）·《月曼清游图·四月曲池荡千》·故宫博物院藏品

　　高 37 厘米，宽 31.8 厘米。

　　陈枚绘清代宫廷妇女的四季生活。茶器置于画面一角。康熙、雍正时内供奉，

内务府员外郎。

图343

清·顾洛（1763—约
1837）·《小倩小影
图》·无锡市博物
馆藏品

绢本设色，高99
厘米，宽39厘米。

顾洛，字西梅，
号禹门，浙江钱塘（今
杭州）人，擅仕女、
山水、花卉。小倩，
万历年间人，杭州豪
公子冯某妾，能诗擅
音律，为大户不容被
置孤山别墅，抑郁而
死。画面描绘小倩庭
院，伏案读诗，小杯
盛茶，一副恬淡孤寂
之感。

图 344

清·改琦（1773—1828）·《仕女图》·广东省博物馆藏品

原画高 28 厘米，宽 39.5 厘米。

图 345

清·张宗苍·《弘历抚琴图》

　　古树下，一杯清茶，一部古书，一张古琴，石床石椅。没有佩剑持戟的武士，没有衣着华丽的仕女。画中是一个静谧、淡泊、儒雅的圣贤。抹去帝王的神圣光环，呈现在人们面前的是一位茶中达人。这就是茶道文化的修养所在。不需要繁缛的程式和仪式。"一人曰神"的茶道精神，通过乾隆已经阐释得淋漓尽致。

图 346

清·乾隆（1711—1799）·《竹炉山房图》

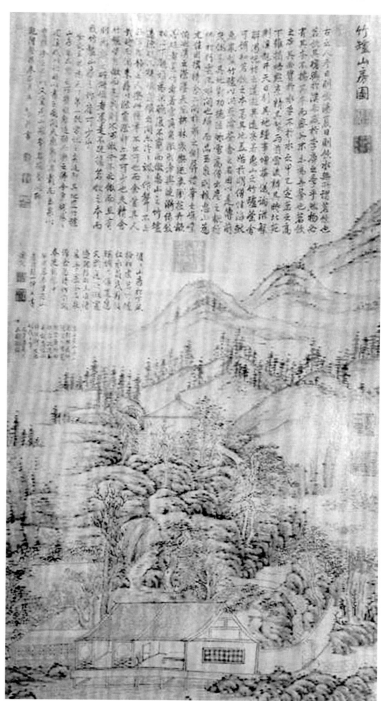

　　1753 年御制，在题记上乾隆写道，茗饮兴于汉而盛于李唐。这将明代人"茶兴于唐盛于宋"的说法向前推进了数百年。他的说法可能更接近历史。因为东汉王褒《僮约》有"烹茶尽具，武阳买茶"之条约。乾隆将历年题诗揭于山房楣楣。

图 347

清 · 金廷标（？—1767） · 《品泉图》

　　金廷标字士揆，乌程（今浙江湖州）人，善人物，兼花卉、山水，白描尤工，亦能界画。清高宗乾隆二十五年（1760）第三次南巡时，进白描罗汉册，受到乾隆帝赏识，命入内廷供奉。

　　此图描绘一文士在皓月之下，坐在溪流边的垂曲树干正在把盏品茗，状态宁静悠闲，一童蹲踞溪中石上用长柄勺取水倒入罐中，另一童在方形竹筐篮中的炉中燃炭，三人的汲水、备茶、啜茗动作，恰恰自然地构成了一幅汲水品茶的连环图画。画面上明月高挂，清风月影，品茗赏景，十分自在。画上的烹茶道具有竹炉、茶壶、四层提篮（挑盒）、水罐、水勺、茗碗等等，斑竹茶炉四边皆绑提带，四层提篮内可容烹茶需要诸品，如茶叶、炭火等等，可以想见图上的这套茶器就是外出旅行用的。品泉，实际是以茶试水的意思。

　　本幅山水人物浅设色，笔墨精练，人物清秀，圆圈转折遒劲，皆与记载中的"折芦描"相近。

图 348

清·金廷标·《仙舟笛韵》·台北故宫博物院藏品

　　涧水默默，笛声悠悠，画家把理想的茶道意境铺展
于笔下。

图 349

清·傅衡·《同甲长春图》

　　傅衡，生卒年不详，同治六年（1867）进士，贵阳人。所绘应为清代进士茶宴情形。

图 350

清·潘承桂·《琴瑟和鸣图》

图 351

清·潘承桂·《芦湖赏月图》

　　潘承桂，生卒年代不详。绢本设色纨扇。作于 1896 年。因其内容与金廷标上图意
趣相近，故排列于此。《琴瑟和鸣图》画中题款："今日天气佳，吹箫与弹琴。"《芦
湖赏月图》款题："丙申新秋　写奉鲁泉大兄大人雅教，子秋弟潘承桂时于寓昭阳客村。"

图 352

清·苏六朋（1791—1862）·《游僧归晚图》

　　表现的是作者青年时期求学时所住广东罗浮山宝积寺，晚归僧人可能是他的老师尚德大和尚或师叔辈。画中题诗"一僧归得晚，云气满袈裟"。款题"道光己酉（1825）三月晦日，将还羊城，为洁大师作。怎道人六朋"，钤白文长方"枕琴"印。

图 353

清·苏六朋·《秋赏图》

　　品茗赏花为宋以后士人所崇尚。《秋赏图》题款："道光乙酉（1845）秋初怎道人六朋"。
主人坐于石凳。石桌上放置一盆花，同时还有紫砂壶、白瓷杯。另有两人抬着大花盆走来。

　　中国茶人自宋代以来，将插花挂画与茶艺结合，使茶道成为一种多种艺能的结合。图中
是品茗，赏花。

图 354

清·王素（1794—1877）·《高士论诗图》·西安博物院藏品

从高提梁紫砂壶直接在火上煮的情形看，很可能就是煎煮云南普洱茶或湖南安化黑茶或陕西茯茶之类。乾陵皇帝有诗云："独有普洱号金刚，清标未足夸雀舌。点成一碗金茎露，品泉陆羽应惭拙。"这里的"点"实则指的是"瀹"，也就是"烹"，是"煮饮"与《红楼梦》六十三回"寿怡红群芳开夜宴"中"该焖些普洱茶喝""焖了一茶缸女儿茶，已喝过两碗了"都是文火煮熟的意思。（黄桂枢《乾陵普洱茶诗考辨》，《云南日报》2007 年 11 月 29 日第 11 版）

图 355

清·费丹旭（1802—1850）·《姚燮纤绮图》·北京故宫博物院藏品

纸本设色，高 31 厘米，宽 128.6 厘米。

费丹旭，字子苕，号晓楼，晚号偶翁，又号环渚生，湖州人。工写照，如镜取影，为人写真留照，名盛一时。其写真，不是单纯的画人物半身或全身像，而是将人物置于一定的场景之中，如《为楚江画像》画高山流水，楚江于松下石旁拥琴而坐，观飞瀑，听泉声，童子在旁煮茶以助幽情。右上款："戊申峡四月，楚江上公将军命写。小楼费丹旭。"

《姚燮纤绮图》绘姚燮与家中诸侍姬真实的生活情景。姚燮趺坐于蒲团上，前置树墩茶床，茶壶、茶杯、香炉等一应俱全，姚燮跂坐于铺地园毯，表情怡然微笑，仕女们或解书囊，或展卷，或抱琴，或捧剑，仪态各异。时在柳影婆娑，桃红绽开的清明前后。此景不禁使人随口吟诵出"桃花未尽开菜花，夹岸黄金照落霞。自昔关南春独早，清明已煮紫阳茶"的诗句。该诗为道光时兴安（安康）知府叶世倬所作。此画虽为江南风物，但清明汲泉品新，

则南北无别焉。图名为"纤绮"，旨在刻画"罗绮丛中悟此身"的禅学意境以及"鬓丝禅榻心暗伤，年华似水不可恃"的感触。画面右上款"己亥秋日为梅伯仁兄属写。晓楼丹旭"。钤朱文印"子苕"。当作于 1839 年秋。姚燮（1805—1864），晚清文学家、画家。字梅伯，号复庄，又号大梅山民等，镇海（宁波）人。道光年间举人，治学广涉经史子集释道，工诗画，尤擅人物、梅花。著有《今乐考证》《大梅山馆集》《疏影楼词》等。

子苕所绘，仕女为长，广涉文士将军，且多以茶事入画，为千余年来湖州茶事薪火相续的写照。从他的画，我们再次感到，中国茶饮，中国茶道文化来源于鲜活多彩的人间生活和修养禅对的宗教实践，生活艺术的灵动曼妙与宗教观想的宁静神秘相得益彰。对于如何恢复中华茶道，我们应更多地总结历史，也要关照生活。

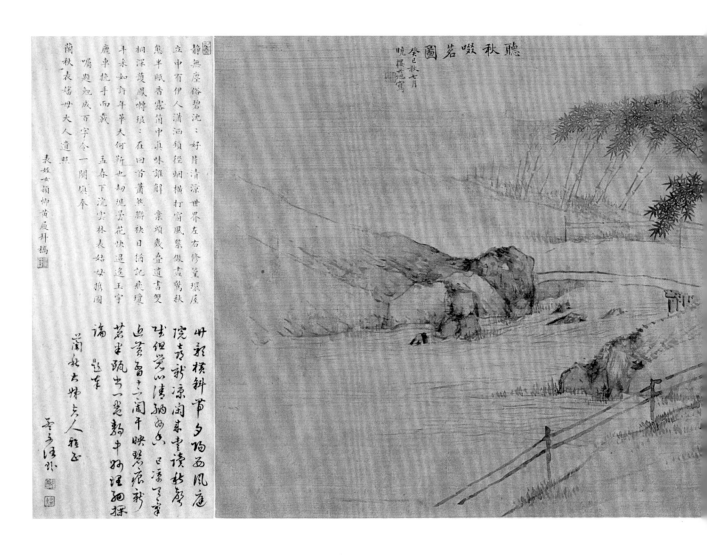

图 356

清·费丹旭·《听秋啜茗图》·浙江省博物馆藏品

高 43.9 厘米，宽 101.6 厘米。

画作右下角题记：

"绿烟轻飏碧云阴，红玉花瓷试浅斟。清绝襟怀谁得似，教儿领略读书心。惆怅春波送远行，吟怀怅触繫离情。龙团品到应思我，独倚秋窗听碧声。碧声山馆 君修四弟家中读书处。

兰秋四妹将赴闽中，濒行出玉照嘱题离怀，结辖不能成章也，口占绝句二首即以赠别伯孙许延礽藁"

钤白文"延礽"朱印，朱文"伯孙"印。

画作左下角题记：

"草影横斜带夕阳，西风庭院喜新凉。闲来爱读新声赋，但觉心清纳妙香。已凉天气近黄昏，十之阑干映碧痕，新茗半瓯出一卷，静中妙理细探论。题奉

兰秋大姊夫人雅正

孟文任珍"

钤白文"任珍印"朱印。朱文"孟文"印。

画芯题名："听秋啜茗图 癸巳秋七月晓楼丹旭写"。

画左上角题款：

"静无尘俗碧沈沈，好片清凉世界，左右修篁环屋，立中有伊人潇洒，锁径烟横，打窗风紧，做尽惊秋态。半瓯香露，个中真味谁解，案头几叠遗书，双桐深护，凤嗦琅琅在回首，萧然联袂日，犹记飞琼丰采，如许年华，天何靳也，劫现昙花快迢遥，玉宇鹿车，挽手而载。

孟春下 。浣云林表姑母携图，嘱题勉成百字令一阕填奉

兰秋表婶母大人遗照　表侄女颖卿黄履拜稿"

钤白文"黄履"朱印。

画作右上角题记：

"修竹影濛濛，翠减幽丛。啼鹃几染泪痕红，难热返魂香一炷。空认芳容，杯棬泽犹浓，画阁尘封，云軿同住碧城中，日落平台携手望，依旧春风。浪淘沙

兰秋夫人遗照　宛生邵懿恒题"

钤白文"宛生"朱印。

读画可以知道，此画是道光十三年（1833）费丹旭32岁时，为许兰秋所作。许兰秋将赴闽中，濒行，拿出费丹旭的画，让哥哥许延礽与任珍（字伯年）题字留念，以表达结缡难解的离别之情。大约兰秋已经对自己的身体状况有所觉察，因而特别留恋兄弟及朋友在一起的日子。我们不太明了赴闽中的原因，是经商，还是做官？但一定是一次远行。我们也不清楚兰秋的离世时间，但比较年轻——黄履以昙花劫来比喻。这幅画后被黄履表姑母浣云林拿来，让她题字。表姑母浣云林与表婶兰秋应是妯娌关系。任孟文，与许延礽、许兰秋为朋友关系。

画中主要人物为兰秋在家中碧声馆，读书、品茗与教育儿子读书的真实场景。兰秋儿子大约十岁左右。

画中梧桐、修竹、流水、石桥、瓶花、玻璃灯罩、瓷坐墩等，布置得恰到好处，人物宁静而和谐。画面整体给人一种舒缓宁静的感觉。"新茗半瓯出一卷，静中妙理细探论。"一句也是清代诗文中，关于探究茶道妙理的不多见的文字。邵懿恒、黄履作为晚辈，在兰秋过世后，应邀题画作诗，表达一种崇敬与悼念。

图 357

清·汪承霈（？—1805）·《群仙集祝图》

　　是对唐宋《斗茶图》的改绘，增加妇女、小孩。对
茶文化史研究有一定参考价值。

图 358

清·袁耀（？—1788）·《潇湘烟雨图轴》·安徽博物院藏品

　　画高 155.8 厘米，宽 58.5 厘米。与米家画舡意境相似，但内容更多。

　　袁耀，康熙后期扬州人。与其伯父袁江深得宋元精髓，工楼阁界画，号"二袁"。画中主体内容一船二人，一人披蓑衣打鱼，一人品茶看书。题款"潇湘烟雨冷崖耀又笔"。钤白文朱方印"袁耀"。

图 359

清·万上遴（1739—?）·《渔乐图》·安徽博物院藏品

万上遴，字殿卿，号辋岗，江西分宜人，曾任清宫画院待诏。本图《渔乐图》有题款"丙子桂秋辋岗上遴"。钤"上遴之印"白文朱印，"辋岗画记"朱文印。丙子在他的生平中有可能经历两个，1756 年，1816 年。如果按照 75 岁虚龄计算，他的出生年龄可能在 1741 年之后。画中为渔民在湖上打鱼、煮茶、备炊、吹笛等活动。饮茶深入江南农民生活的每一个角落。

图 360

清·沙馥（1831—1906）·《携琴访友图》·西安美术学院藏品

　　作者 1887 年作，携琴访友，靠树倚石，书童在后，茶不可缺——
在画外，友人处。

图 361

清·沙馥·《琼筵飞觞图》·西安美术学院藏品

高 58.5 厘米，宽 91.3 厘米。

或以为其取意为李白《春夜宴从弟桃园序》之意："开琼筵以坐花，飞羽觞而醉月。"杏园之内，在石山之旁，一帮文人墨客，以文相会，茶酒助欢。举杯仰望天空者，当为李太白，把酒邀月，快意诗赋从他的口中飞流而下，旁边四人，为聚一起，赶紧奋笔记录。迟到的文士带着书童挂杖缓缓而行，似乎早已融进诗人悠远高妙的诗情画意之中。宽敞的石桌山有直筒酒杯，也有带执手的茶杯。杏树桃树的粉色花瓣已经转为白色，春天已经悄然到了人们的身边。

图 362

清·钱慧安（1833—1911）·《烹茶洗砚图》·上海博物馆藏品

纸本设色，高 62.1 厘米，宽 59.2 厘米 。

钱慧安，字吉生，号清溪樵子。宝山（今属上海市）人。幼从事丹青，擅工笔画，以人物、仕女为专长。间作花卉、山水，均能自出机杼，不落前人窠臼。《烹茶洗砚图》是钱慧安为其友文舟所作的肖像。此画在两株虬曲的松树下，有傍石而建的水榭，一中年男子倚栏而坐。榭内琴桌上置有茶具、书函，一侍童在水边涤砚，数条金鱼正游向砚前；

另一侍童拿着蒲扇，对小炉扇风烹茶。人物线条尖细挺劲，转折硬健，师法陈洪绶而不受所围，其技法已臻纯熟，仪容娴雅，设色清淡，为清末海上画派的风格。这是同治十年（1871），作者 39 岁时，为友人文舟所作的肖像画。画面左上角篆书自题"烹茶洗砚"，另行行书："辛未新秋凉风渐至，爽气宜人，适文舟尊兄大人属布是图，聊以报命，即希正之。清溪樵子钱慧安并记。"下钤"吉生"朱文圆印。

图 363

清·钱慧安·《人物花卉四条屏·柳下读书》·西安考古所藏品

　　画左上题字："窗前垂柳如人立，长日低头听读书壬午嘉平之吉仿新罗山人本清溪樵子，钱慧安。"画中两少儿前面各一本打开的书本。一个端坐而低头，另一则仰望教书先生，接受老师考问。桌上，一戒尺，一紫砂壶，一青花瓷杯。一水盂，一砚台。屋外一妇人，侧身站在墙角，偷偷观察他们的教与学。从装饰和发型看，很可能是孩子的母亲。

图 364

清·任薰（1835—1893）·《大梅诗意图》之一

纳凉，吹笛，船上有方桌、白釉茶盏、茶灶、贮水缸。

图 365

清·任熊（1823—1857）·《人物》册页·天津博物馆藏品

　　画高 28 厘米，宽 36.8 厘米。

　　任熊，字渭长，一字湘浦，号不舍，浙江萧山人，后寓居苏州、上海，以卖画为生。浪
迹到宁波，师从姚燮，得其宋人笔法。熊与弟任薰、儿子任预、侄任颐合称"海上四任"，
又与朱熊、张熊合称"沪上三熊"。《人物》册页之四以唐代道士玄真子张志和命樵青竹里
煎茶的故事为题材而作。修竹、石案石堆，湖水，前一童子持扇对风炉烹茶。画左下角钤朱
文"阜长"印。

图 366 为真然（莲溪）《淮阴垂钓》，图 367 为莲溪《临泉调琴图》，二者均为绢本设色扇面画。两幅画看起来，与煮水饮茶没有直接关系。但我们认为，就广义的中国茶道文化而言，形而下的烹茶尽具，实际与抚琴垂钓，有异曲同工之妙。垂钓与候汤品茗都是以宁静恬淡的形态达到心绪的平和开阔。而琴棋书画作为茶人四艺，与品茗自有天缘。就精神层面而言，都在追求天地间的至真至妙。古往今来的高妙音乐是通过乐器来描摹天籁之音。而在高山之腰，修竹之下，泉瀑之旁，琴者拨指弄弦，是对泉水的飞流与撞击，风对竹林的推掀裹挟，鸟儿啄木，蚊虫鸣叫，水滴对树叶的滴落拍打等等，所进行的描摹和顺应。这种种声音融合在一起的合奏，便是人与自然，天地之间的对话与和谐。无疑，诗书画琴（乐）与茶有高度的融合性。宁静、淡泊、中和，乃茶道之主色调。即使喝了卢仝的第七碗茶，也无非就是羽化成仙，翱翔于天地间，朝饮甘露夜邀月。并不追求力拔山兮气如虹的扬厉与狰狞！

图 366

清·莲溪（1816—1884）·《淮阴垂钓》·西安美术学院藏品

图 367

清·莲溪·《临泉调琴图》·西安美术学院藏品

图 368

清·倪田（1855—1919）·《钟馗仕女图》·徐悲鸿纪念馆藏品

　　纸本设色。倪田，初名宝田，字墨耕，江苏扬州人。侨居上海，画初学王小某（慬），人物、仕女、古佛像等皆佳。尤善画马。因喜爱任伯年画，故作画亦受伯年画法影响。设色花卉、人物，风格近似伯年，古拙绝俗，意境清新。《钟馗仕女图》将钟馗的粗率勇猛与仕女的沉秀纤柔构成一种诙谐的对照，画面显得生动而有趣。

图 369

清·叶芳林·《九日行庵文燕图》（局部）·美国克里夫兰艺术博物馆藏品

绢本设色。高 31.7 厘米，宽 201 厘米 。

叶芳林，生卒年不详，字震初。吴县（今属江苏）洞庭山人。工写照。此画描绘了马氏兄弟在行庵会友雅集的情景。行庵在扬州天宁寺西隅，马氏兄弟所购。文士聚会，或品茗论道，无关日月短长。或观花赏画，妙趣天成；或吟诗赋景，精神深奥。案上水盂、茶壶、托盏映现衬托出他们散淡自然的意趣。

图 370

清·吴晨·《高隐图》·西安博物院藏品

　　吴晨，生卒、行状不详。左图为其《山水册》五页中的一幅《高隐图》。右图为其右上角题诗一
首。无款识。钤朱文印"漱芳"诗云："若其日，谈云漱晚，香风细竹几斜横。藤床半据，蕉影侵帘，
桐阴覆地，书卷初安，茶烟欲沸，聊盘礴以怡情，暂安舒而鲜虑。录《幽窗赋》一则　补徐吏部法。"

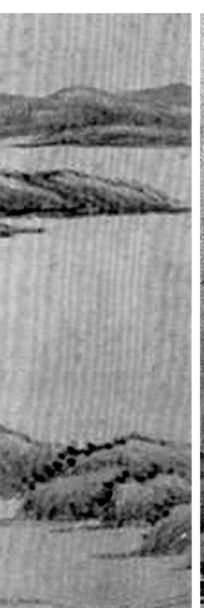

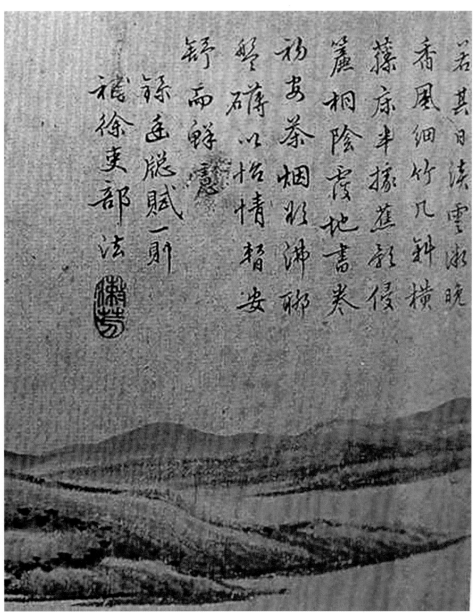

若其日落云淡晓
香风细竹几斜横
藤床半据蕉衫侵
簾桐阴霞地书卷
声砖以怡情智安
杨安茶烟松聊
舒而鲜蕙

禄匜胧赋一
裨徐吏部法

吴晨·《高隐图》（局部）

图 371

清·佚名·《十不闲图》·天津市博物馆藏品

画高 65 厘米，宽 115 厘米。

《十不闲》是满族弟子的一种说唱词，以"万福骈臻""瑞谷丰登"为主题。此画描绘的是儿童们在节日里吹拉弹唱，玩打"十不闲"的场面。右边的石茶几上一只黄釉茶壶，几个外蓝内白釉的茶杯，表现出渴时饮茶、边玩边喝茶的情景。茶，竟也是满族贵胄儿童的喜爱！

图 372

清·佚名·《柳岸货鱼图》
（局部）·西安考古所藏品

纸本。全幅高 125.5 厘米，
宽 66.5 厘米。

画中两木船并列靠岸。老
人在一船上举起小孩脸顶其面
颊，小孩左手紧攥一只拨浪鼓。
近处船上一妇女专心致志地用
紫砂壶在炉子上煮茶。船上有
白瓷水盂、勺子。白瓷瓶子里
黄花绿叶。炉膛敞大，火焰旺盛。
岸上卖鱼者右手提秤，左手移
动秤锤，双腿挽起裤管，赤脚，
头戴草帽。买鱼者白衣蓝布圆
口鞋，戴草帽，腰上系黑腰带。
眼睛盯着秤星。另一位秃顶白
胡子老者弯腰与一白衣小孩挑
拣大鱼篓子中的鱼。右手扶着
篓子边，左手食指伸出，与卖
鱼者说话。而卖鱼人眼睛也盯
着秤星。从绿草和柳树色彩看，
可能是秋天景象。整个图画给
人一种恬淡的渔乡景象。画中
头戴巾帕穿白衣的妇女手持白
色大茶碗，静静地听音侯汤。

图 373

清·任伯年（1840—1896）·《灯下织机》·北京故宫博物院藏品

　　原画高 24 厘米，宽 38.2 厘米。

　　任伯年，名颐，字伯年，浙江萧山人。画中题款："光绪戊子任颐"，知其作于光绪十四年（1888）。
画中为茅屋之下，一中年妇女正在灯下穿梭织布情形。左边木桌上四腿铁架油灯，一紫砂执壶，一白
瓷直筒茶杯。这与属于阳春白雪的茶道文化不同，属于"柴米油盐酱醋茶"的茶。茶在不同环境中有
不同的物质和精神意义。茶的物质意义是茶文化的基础。

图 374

清·张筠·《说书纳凉》·西安美术学院藏品

　　1851 年作，纸本设色扇面，高 19.1 厘米，宽 52.2 厘米。

　　张筠，生卒年不详，浙江嘉兴人。1861 年作。画中为表现浓荫将树冠、树窝和竹叶都涂为黛色和黑色。画面焦点为说书人，他靠椅倚桌，左手执牙板，右手敲击手鼓，绘声绘色；听书者，男女老幼，听得认认真真、津津有味。方桌之上白瓷执壶与紫砂壶各一，说书人前一白瓷杯。卖个关子，抖个包袱，抿一口清茶，别有一番快意。听书人，仔细听故事，想哲理，品新茶。袒胸哺乳的夫人听书入迷忘了变换喂奶的姿势，站着听书的人全然不知道小狗在他簸箕里偷食食物。除了私塾学堂和社戏，喝茶听书大约成为乡村中老少咸宜的最高雅的文化活动形式。

图 375

清·吉朝·《灵山积玉图》·西安美术学院藏品

　　作者生卒年不详。此画作于 1899 年，高 136.7 厘米，宽 73.7 厘米。

吉朝·《灵山积玉图》（局部）

图 376

清·潘振镛（1852—1921）·《天街夜色》·西安美术学院藏品

画高 99.1 厘米，宽 39.2 厘米。

潘振镛，浙江嘉兴人，字承伯，号亚笙、雅声，自称壶琴主，晚署讷钝老人，钝叟，钝老人。此画作于 1910 年。题诗："天街夜色凉如水，坐看牵牛织女星。"款署："庚戌蒲夏用唐解元笔意写唐人诗　潘振镛"，下钤"潘振镛"白文方印，"雅声"朱文方印。月高夜静，小树小草有了精神，桐叶深沉。摇扇摧风，静享寂静和一丝清凉。三足香炉的烟气已经消散，紫砂壶和玻璃杯尚有茶香淡淡。七月流火，织女牛郎鹊桥会。郎未在，不知佳期长短。正寂寞，有娇儿蹒跚来伴，相思更添。唐解元乃唐寅，唐伯虎也，以仕女画见长，善于表达仕女的幽怨伤感。而潘振镛的这幅画则给人以平和舒畅的美感。

图 377

清·任预（1853—1901）·《残月晓风图》·日本桥本末吉藏

高 102.1 厘米，宽 39.6 厘米。

题材取自宋柳永《雨霖铃》。一叶孤舟泊清秋，彻夜情话到残晓。去也，归也，无定期。红烛已暗，残月将去，杯中茶，万千离别意。作者浙江萧山人，为任熊之子。"立凡"印，题款表明，画作于光绪二十三年（1897）。

"和、雅、敬、健"中华茶

Harmony, Elegance, Piety, and Persistence Seen in Chinese Tea

一、中华茶道文化

中国有否茶道，这已经不是学术问题，而是如何实际操作、如何对外传播的问题。

自东晋杜育《荈赋》开始，茶就进入中国人的精神文化视野，公元780年前后，陆羽《茶经》问世。"茶道"一词最早出现在陆羽好友诗僧皎然的《饮茶歌诮崔石使君》的末尾两句："孰知茶道全尔真，唯有丹丘得如此。"封演与陆羽、皎然也是同时代人，他的《封氏闻见记》认为"楚人陆鸿渐为茶论……于是茶道大行"。

茶道，是在一定的环境气氛中，以制茶、饮茶、烹茶、点茶为核心，通过一定的语言、身体动作、器具、装饰和环境表达一定思想感情，具有一定时代性和民族性的综合文化活动形式。

A. Chinese Tea Culture

Whether there once had Chinese tea ceremony or not has not been an academic issue. The actual focus is how to practice Chinese tea ceremony, make more communications with foreign countries and regions via it.

Tea culture gradually became part of Chinese spiritual life since the time when Du Yu of the Eastern Jin Dynasty wrote "Ode to Tea Planting and Drinking" which is the earliest Chinese literary work on tea. *The Classic of Tea* by Lu Yu came out around 780 of the Tang Dynasty. The word "Tea ceremony" first appeared at the end of the poem "Tea Drinking Song as a Satire towards Cui Shishi" by Jiaoran, a monk poet and also a good friend of Lu Yu: "Who knows one can attain the Tao by drinking tea? Who knows one can attain the true Tao fully? Maybe only the legendary immortal Dan Qiuzi knows." Feng Yan, who was a contemporary of Lu Yu and Jiaoran of the Tang Dynasty expressed his own idea in *What Have Been Seen and Saw By Feng Yan*, "Lu Hongjian (namely Lu Yu), a Chu person wrote a book discussing issues on tea... and the tea ceremony gradually became popular."

Chinese tea ceremony is a form of comprehensive cultural activities conveying certain thoughts and feelings through typical presentations, body movements, appliances, decorations

我们主张综合文化概念的理由很简单。第一，茶为饮用而生产的，茶是健康饮料。第二，茶文化的核心是品茶道。第三，不论历史上还是现在，虽然每一次品茶过程各不相同，但都包含了物质和精神两个方面；每次茶会的主题因缘各不相同，但都有一个情感的交流、思想的碰撞与精神的升华。第四，作为茶道文化的元典，陆羽《茶经》包括了与茶有关的方方面面，他将持续千年的茶事活动上升到文化建设、个人修养的高度。第五，茶道审美决定了茶叶茶具设计与生产、茶室的布局与装饰。在唐代，卢仝说"天子不尝阳羡茶，百草不敢先开花"；而到了北宋，建茶建盏号为天下第一；乾隆下江南，唯龙井是其最爱。这是烹茶道、点茶道和散茶冲泡三种茶道模式下，三个时代的天下第一名茶！第六，求真、自然、精妙的古代茶道审美与今天人们对健康茶叶的要求不仅一致，而且现代化工艺设计与器具的制作更有益于茶道审美在茶具制作和茶汤出色方面的展示延伸。第七，沿着陆羽当年文化建设和提高个人修养的思路，重新确立茶道思想的核心概念是当

and environment of which the core contents are making tea, brewing tea and drinking tea in a certain atmosphere. Chinese tea ceremony possesses contemporaneity and nationality.

The reasons for us to advocate the concept of Chinese tea ceremony being a form of comprehensive cultural activities are as follows:

1. Tea is a healthy drink and the purpose of tea production is for drinking; tea ceremony is the core content of Chinese tea culture.

2. Although the tea drinking processes are not identical in different periods of history, its material and spiritual connotations are both there.

3. The cause and theme of the tea party vary from time to time, but there are always emotional communication, the collision of thoughts and the sublimation of the spirit.

4. *The Classic of Tea* as an initial classical book on Chinese tea culture includes all tea−related aspects discussing one−thousand−year old tea activity at a much higher level of cultural construction and personal accomplishment.

5. Aesthetic criterion of tea ceremony exerts a decisive influence on the tea production, design and manufacture of the tea set, layout and decoration of tea−drinking room. In the Tang Dynasty, Lu Tong said, "All kinds of flowers dare not to bloom before the emperor tastes the Yangxian tea." The Jian tea and Jianzhou kiln calyx are second to none in the Northern Song Dynasty. The Emperor Qianlong of the Qing Dynasty loved the Longjing tea the most while he made inspection tours in southern China. Tea cooking, dust−tea brewing, loose tea brewing were three major tea−making methods during the Tang Dynasty, the Northern Song Dynasty, and the Qing Dynasty. The Yangxian tea, the Jian tea and the Longjing tea were the most famous types at the

代文化建设的需要。

二、应重新确立茶道文化 的核心概念

近百年来，特别近三十年来，中国当代茶人从庄晚芳到陈文华，曾经企图借鉴日本茶道的"和、敬、清、寂"来概括中国茶道文化的理念，至目前为止，还没有形成一个大家共同认可的茶道思想的核心概念。这是一个亟待解决的学术问题，也是茶道文化实践的当下需要。

中国古代茶人的茶道思想理念归纳如下。

皎然：清高、爽朗、全尔真。

陆羽：茶宜精行俭德之人。

晚唐苏廙著《十六汤品》："中庸""守一""和谐""完善"（赖功欧总结）。

唐代诗人韦应物、齐己、郑遨、陆龟蒙、杜牧：清高、幽静、空灵。

晚唐刘利贞：以茶利礼仁，以茶表敬意，以茶行道，以茶可雅致。

宋徽宗赵佶《大观茶论》：祛襟涤滞，致清导和，中淡闲洁，韵高致静。

明代朱权《茶谱》：茶之为物，可以助诗兴而云山顿色，可以伏睡

times.

6. The pursuit of truth, harmony with nature and delicacy as the key points of ancient Chinese tea ceremony aesthetic criterion is in accordance with people's wish to obtain health by drinking tea. The modern design of arts, crafts, and the manufacturing of the tea sets are favorable for the conveying and presenting ancient Chinese tea ceremony aesthetic criterion which reflected by the high quality tea set and tea itself.

7. Tea activity has the function of personal accomplishment and cultural construction according to Lu Yu's view point. We need to reconfirm it because that is the core concept of tea ceremony, and also make it to serve the contemporary cultural construction.

B. Re-establishing the Core Concept of Chinese Tea Culture

Over the last century especially in the past nearly 30 years, China's contemporary tea-man including Zhuang Wanfang and Chen Wenhua attempted to summarize the concept of Chinese tea culture by taking that of Japanese tea culture – harmony, piety, purity, and tranquility as a reference, but no generally-accepted core concept of Chinese tea ceremony has taken shape so far. This is an academic problem needed to be solved urgently also a current need for tea ceremony practice.

Thoughts and Ideas of tea-men on tea culture in ancient Chinamay be summarized as follows.

Jiaoran: Self-contained, cheerful, whole-heartedly, and frank.

Lu Yu: Tea as a drink was suitable for virtuous person.

Su Yi of the Late Tang Dynasty wrote in his book *Sixteen*

魔而天地忘形，可以倍清谈而万象惊寒，茶之功大矣……探虚玄而参造化，清心神而出尘表。

当代台湾学者范增平、周渝和大陆学者庄晚芳、陈文华、余悦、赵天相、周国富、程启坤、姚国坤、丁以寿、丁文、王铃、陈香白、林治及马嘉善等17人次提出不同的中华茶道思想的核心概念，统计后可以看到：和8、敬5、美5、雅2、清3、健1、静4、性1、怡2、真2、洁1、悟2、伦1、俭2、理1、正1、圆1、融1、中庸和谐1、中和1、礼2、廉1、天人合一2、大用1、大悲1、道1、尝1、自然1、清雅1、明伦1。

可以看出，在中华茶道思想核心概念上，人们在"和""敬""雅""美"外，还有更多概括，还没有形成共识。

三、"和、雅、敬、健"应为最高结晶

毫无疑问，以上茶学专家的提法既有实践总结的倾向，又有理论提炼的旨趣。但是有几个不足值得改进：(1)作为一种文化观念或文化理论的层次不清；(2)重复；(3)作为一个文化系统，内容缺失；(4)缺乏理论逻辑而偏重于茶文化感性认识。

Tea-making Methods: Moderatamente, single-mindedness, harmonious, and perfection (summarized by Lai Gong'ou).

Poets of the Tang Dynasty such as Wei Yingwu, Qi Ji, Zheng Ao, Lu Guimeng and Du Mu: Self-contained, peaceful and intangible.

Liu Lizhen of the Late Tang Dynasty: Tea showed rite, benevolence, piety, Taoist belief, and elegance.

Zhao Ji namely Emperor Huizong of the Southern Song Dynasty wrote in his book *Discussion on Tea*: Tea helped people to get clear and pure mind, harmonious relation, leisurely and comfortable state, and rhythmic tranquility.

In Zhu Quan's *Tea Manual*: Tea could inspire the poet's imagination, chase away the desire of sleep, create pleasant atmosphere for idle talk, explore nature in tranquility and clear one's mind to think like an immortal.

Contemporary Taiwan scholars such as Fan Zengping and Zhou Yu, mainland scholars such as Zhuang Wanfang, Chen Wenhua, Yu Yue, Zhao Tianxiang, Zhou Fuguo, Cheng Qikun, Yao Guokun, Ding Yi, Ding Wen, Wang Ling, Chen Xiangbai, Lin Zhi, and Ma Jiashan, 17 of them altogether put forward different ideas on the core concept of Chinese tea culture, the related statistics show as the following: 8 of them take Harmony as the core concept of Chinese tea culture, 5 piety, 5 Graceful, 2 Elegance, 3 Purity, 1 Persistence, 4 Tranquility, 1 Personality, 2 Joyful, 2 Frank, 1 Clean, 2 Insight, 1 Moral, 2 Thrifty, 1 Rational, 1 Righteous, 1 Flexible, 1 Open-minded, 1 Moderatamente, 1 Neutralization, 2 Rite, 1 Honest and upright, 2 Unity of human and nature, 1 Useful, 1 Merciful, 1 Law, 1 Experience, 1 Natural, 1 Refined, 1 Clearly Understood Ethics.

It is obvious that no consensus has been reached in terms of the core concept of Chinese tea culture besides of the initial

要在中华茶道思想的核心概念上形成共识，我们认为，应该具有四个标准或功能：在哲学层面能够充分反映中国传统的价值观和审美观，并契合时代特征；在物质和环境层面上能够揭示作为文化载体的茶的特性和功能；在茶道文化的社会意义层面上应该具有鲜明的时代特征，或者符合中华文明的发展方向和轨迹；在思想内容上基本涵盖以上古今茶人的茶道思想主张。

基于以上四个方面的考虑，我们试提出"和、雅、敬、健"四个概念。

哲学精神与价值观层面和审美层面上的总体要求：和。包含上述的"和、中和、中庸、天人合一、全尔真、圆、融"。

大家对"和"这一中国文化精神的根本要义一致坚信不疑。如果把"和"作为茶道的宏观主旨，是中华文化的最高理念和价值观，是精神文化的核心元素，统摄社会和人生的终极目标、规范人的行为和审美境界。包含四个层次的概念：大和、和谐、中和与和合。先秦儒学集大成者孔子主张："礼之用，和为贵。先王之道，斯为美；小大由之，有所不行，知和而和，不以

consent in the following aspects: harmony, piety, elegance, and graceful.

C. Harmony, Elegance, Piety, and Persistence Are supposed to Be Outstanding Achievements

There is no doubt that the above-mentioned summarizations by scholars from Taiwan and mainland China have both elements of practice and refining of theory. But there are some deficiencies needed to be improved: 1. No clear methodical sequence as a cultural concept or theory; 2. Repetition; 3. Lack of content as a cultural system; 4. Lack of theoretical logic but mainly focusing on perceptual knowledge of tea culture.

We think that there should be four standards or functions for reaching a consensus on the core concept of Chinese tea culture: 1. Fully reflect traditional Chinese value and aesthetic view, and conform to characteristics of the time at the philosophical level; 2. Reveal the feature and function of tea as a cultural carrier at material and environmental level; 3. Have distinct characteristics of the times or conform to the direction and path of the development of Chinese civilization at the social level; 4. Contain basically the major ancient and modern thoughts of Chinese tea-men on tea culture in terms of ideological content.

I am trying to propose four elements "harmony, elegance, piety, and persistence" as the core concept of Chinese tea culture based on the consideration of the aboving four aspects.

"Harmony" covers the above-mentioned several expressions: harmony, neutralization, moderatamente, unity of human and nature, whole-heartedly and frank, flexible, open-minded in consideration of philosophical spirit, values, and aesthetic demands.

礼节之，亦不可行也。"（《学而》，见《论语集注》）管子、老子、墨子也都尊奉这一价值与审美理念，主张"和""和合"。

此外，我们在这里强调的是，实际上从字义以及"和"的行为体现看，"和""和合"还有另外一个含义："圆融""化合"。《道德经》云："万物负阴而抱阳，冲气以为和。"这里的"和"，包含了阴阳转化与"和合"含义。

在审美上以"中和""中庸"为最高境界。

"雅"，则是主要体现在对茶的物质自然属性和茶道环境、和行为层面的基本要求。涵盖以上"雅、大雅、俭、美、廉、清和、静、洁、怡"，涵盖皎然等的清高、空灵、和幽静等含义。

就精神层面来讲，"雅"应该是精妙、平衡、美好、适度、廉俭得宜，是宁静、不喜亦不悲的淡淡禅悦之美、忘我无羁的超脱自性！况且，无论是煎茶自饮还是瀹茶待客，都是高雅的活动，而一期一会的品茶聚会，更是综合性的高雅活动形式。而这里的"雅"更偏重"清""淡泊""宁静"。这个雅字实际已经蕴含了简

Everyone believes that harmony is the fundamental point of Chinese culture and Chinese spirit. If we regard harmony as the keynote of Chinese tea culture in the macroscopic scale, then we truly understand the highest idea and value of Chinese culture, and the core element of Chinese spirit which has controlled the social and personal ultimate goal, and regulated people's behavior and aesthetic realm as well. It contains four points: harmony at the highest level, harmonious, harmony at the medium level and concordance. Confucius, the Pre-Qin Period Confucianism master, advocated that "Harmony is the most precious element in the application of rites; kings of earlier dynasties have executed their rule according to rites; if we rigidly handle issues, no matter big and small, in the light of harmony and moderatamente, sometimes get nowhere, this is because it won't work if we just try to be harmonious on the basis of harmony without being abstinent in terms of rites."(quoted from "Learning for Knowing" of *Collected Annotations on the Analects of Confucius*) In ancient China, Guanzi, Laozi, Mozi also accepted harmony and concordance as the key point of their ideas on value and aesthetic standard.

In addition, what we emphasize in here is that in fact from the meaning of the word harmony and this-word-related behavior, the connotation of harmony and concordance has another implication: flexible, open-minded, unite Yin and Yang after their mutual transformation. It is written in *Tao Te Ching* that "All things tend to embrace Yang when they have more Yin elements than normal standard and reach a new harmony through adjusting."

Harmony at the medium level or moderatamente is the highest realm from Chinese aesthetic standard point of view.

"Elegance" is the basic requirement for the material nature

约、平淡、矜持不奢、内心清静、情绪和悦这些意蕴。老子说"致虚极，守静笃"。一个贪婪奢靡的人，一个情操低下阳奉阴违的人，谈不上雅。雅，也是我们对茶艺、茶器、茶馆环境的基本要求。

选择茶叶以真贞纯粹、清朗明丽为雅，浑浊滥为残朽。茶艺过程以追求不急不滞为雅，以含蓄圆融为雅，鲁莽单调为粗鄙；茶器使用上平和有序为雅，拮抗混乱为非。在茶汤准备上，以隽永精妙为雅，以过浓过淡为野糙。在奉茶待客时以知礼敬尊为雅，以自大无礼为粗蛮。在自我修养上以宁静淡泊为雅，焦躁贪欲为低俗；清高空灵为雅，以粗陋狭隘为粗卑。

在中国行为文化中，人们讲雅，实际上已经远远涵盖或超越了"美"的层次！"雅"不仅滋养我们的眼耳鼻口舌身，更愉悦我们的心灵与精神，正如《金刚经》所谓"生清净心，不应住声、香、味、触法"。

社会伦理建构和生命价值认同、人生修行方面，我们提倡"敬"。涵盖以上"敬、礼、伦、明伦、大悲、悟"。

任何文化都应该以生命为至尊，为生命着想为生命服务。

of tea itself, environment of tea ceremony and tea-ceremony behavior covering several above-mentioned meanings such as elegance, elegance at the highest level, thrifty, graceful, honest and upright, pure and harmonious, tranquility, clean, joyful, self-contained, intangible and peaceful.

Elegance should be regarded as exquisite, balance, happy, moderate, honest and upright, thrifty, tranquility, not happy not sad or slightly happy and selfless at the spiritual level. Besides, both tea cooking, offering tea to guests to drink and periodical tea party are comprehensive elegant activities. Laozi said: "The emptiness and quietness in my heart are to achieve perfection." A greedy and extravagant person or a vulgar-tasted and ostensibly-obedient person can never be elegant. Elegance is also the basic requirement for tea art, tea ware and tea-house environment.

Good tea leaf should be pure, clear in outline, bright in color, vice versa. Elegant in tea art contains no hurry and no delay, subtle, flexible and open-minded, vulgar in tea art contains tedious, crude and rash; elegant in tea ware use contains peaceful and orderly, vulgar in tea ware use contains antagonistic and chaotic; elegant in tea making contains meaningful and subtle, vulgar in tea making contains too strong taste or tasteless; elegant in offering tea to guests to drink contains ritual and respectful behaviors, vulgar in offering tea contains arrogant and rude manner; elegant in personal accomplishment contains tranquility and indifference, self-contained and intangible, vulgar in personal accomplishment contains impatient and greedy, rough and narrow-minded.

In Chinese people's behavior culture, elegant mentioned by people actually covers or far exceeds the meaning graceful. Elegant behavior not only nourishes our eyes, ears, noses,

敬，敬畏，礼敬。这是一种生命意识：敬畏天、地、人，敬畏文明而厌弃野蛮；崇尚真善美而反对假丑恶。仰羡劲松之高岸更怜爱小草之卑微；叹社会人生之曲折复杂，更理解生命个体之无奈渺小；乐山之刚健爱水之柔弱。因为敬畏，更知道以礼相待人事、珍惜物情、欢喜际遇。人生有顺逆，生命无贵贱。中国茶人敬畏崇高，也善待平凡。如孟子所赞美的"君子莫大乎与人为善"（《孟子·公孙丑章句上》）。

敬，也是儒、释、道和伊斯兰教、基督教在信仰和礼敬主尊、祖先方面共同的心理和行为特征。

茶人修行的目标和顺应现实引领未来而言：健。以此涵盖以上的"健、正、大用、真"。

在当今时代资源日渐短缺、经济发展不平衡导致的冲突加剧、因价值观有差异而引起的互不信任越发鲜明，要实现世界大同，建设永久的和平，世界真正的茶人务须自强不息。从一杯茶开始，来传递和平与敬爱的理念，追求和谐清雅基础上的劲健、刚健、勇健、刚正，追求身心的康健。不仅仅是"晨前命对朝霞，夜后邀陪明月"，更是

and mouths but also makes our heart and soul joyful just as what is written in *Diamond Sutra*: "Have pure heart without indulging in sound, scent, taste, etc., keep the whole universe in mind and preserve tranquility and sincerity in soul."

We advocate "piety" which covers the meaning of the following expressions "politeness, courtesy, ethics, clearly understood ethics, mercy and insight" in the aspects of social ethics construction, life value identification, and personal improvement.

Any culture should take life as the supreme aim, always think for the sake of life and serve life sincerely.

Piety with fear and rite, is a kind of life consciousness; refers to respectful manner with fear towards the heaven, the earth, human being, human civilization, truth, kindness, beauty, noble thoughts and ordinary deeds as well; understands the social life also tininess of the individual, vigor of mountains also weakness of water, despises attitude against barbarian. It is natural for us to treat others equally and politely, cherish positive feelings and favorable or unfavorable turns in life due to this. Chinese tea-man's equal kindness to noble thoughts also ordinary deeds is like the similar description in "the First Section of Records of Gongsun Chou's Words" of *Mencius*: "Being kind to others is the supreme nobility for a gentleman."

Piety with fear and rite is also the common psychological and behavioral characteristics of believers of Confucianism, Buddhism, Taoism, Christianity, and Islam while paying their worship.

We advocate "persistence" in the field of personal improvement for tea-man so as to conform to reality and lead the future. Persistence in here covers the meaning of persistence, righteous, useful, and frank.

Nowadays, conflicts caused by increasing shortage of

高岗历风雨，水火见精神！《周易》曰："天行健，君子以自强不息；地势坤，君子以厚德载物。""乾始能以美利利天下，不言所利。大矣哉！大哉乾乎！刚健中正，纯粹精也。"

易学中的"健"的含义，实际上已经涵盖了"正""正能量"和佛教的"正精进"。与《道德经》中"道法自然"有相近的取喻之意。

这也是我们从哲学精神价值观、物质环境观、生命观和未来观四个方面对中华茶道理念的体悟与把握，兼顾了思维科学上的价值、结构和应变三个维度。

以"和、雅、敬、健"为茶道思想核心概念，不仅在内涵上相互兼容，互为主副，在实际上更是相互统摄与协调。旨在为中华茶立命，凝聚正能量，为世界开太平。

《中华茶道图志》的编纂与出版，旨在为事茶者提供具象的历史茶事资料，为茶道程序恢复、茶器制作、茶馆装饰及服饰家具的恢复提供基本的参考。作为茶文化界关注的盛事，本书得到茶人的襄助，尤其是余悦、丁以寿二位教授的审阅、推荐，在此一并表示感谢！

是为跋。

resources and imbalanced economic development have been intensified, more mistrust caused by the difference in value have occurred. We should constantly strive to become the real tea-man in order to realize the world commonwealth and ever-lasting peace. We should convey the idea of peace and love through tea culture, lead the fashion of pursuing physical and mental fitness represented by forceful, energetic, brave, and righteous based on harmony, purity, and elegance.

Persistence in the Yi-ology covers the meaning of righteous, positive energy, also the meaning of positive enhancement from Buddhist point of view, similar to the meaning "Learn from Nature" in view of Taoism from *Tao Te Ching*.

This is our understanding and attempt of defining ideas on Chinese tea culture from the view of philosophical spiritual values, material environment, life and future, meanwhile considering the three dimension of scientific thinking— the value, structure, and strain.

The core concept of Chinese tea culture "harmony, elegance, piety, and persistence" are compatible with each other in the connotation, coordinating with each other in practice. We wish to do something in such way for Chinese tea culture, condensing the positive energy and making peace for this world if possible.

The publication of *Graphical History of Chinese Tea Ceremony* aims at providing tea-man representational data about the history of Chinese tea activity simultaneously a basic reference for the resuming of tea ceremony, tea ware manufacture, tea house decoration, tea-ceremony-related costumes and furniture.

This is the postscript.